*Semi-empirical
self-consistent-field
molecular orbital theory
of molecules*

Semi-empirical self-consistent-field molecular orbital theory of molecules

J. N. Murrell and A. J. Harget
School of Molecular Sciences,
University of Sussex

WILEY–INTERSCIENCE

a division of John Wiley & Sons Ltd
London–New York–Sydney–Toronto

Library of Congress Catalog Card No. 71-172470

ISBN 0 471 62680 5

Printed in Great Britain by The Universities Press,
Belfast, Northern Ireland.

Preface

In the summer of 1970, one of us (J. N. M.) had the opportunity of presenting a course of lectures on semi-empirical SCF–MO theory at the University of Nebraska, and at San Diego State College. In preparing these lectures we were particularly struck by the difference in philosophy between some of the principal workers in the field, and by the wide variation in the methods used to choose the parameters that enter these theories. Moreover the theories have changed rapidly over the last few years, and it must be confusing for the non-specialist to sort his way through the maze of initials ranging from CNDO/1 to MINDO/2. We hope in this book to play the role of pathfinders in this respect.

In writing the book we have in mind, primarily, the experimentalist who is interested in using theoretical methods to predict the results of possible experiments and to analyse his results. At the present time computer programs are readily available for many of the theories which we shall describe and the main problem that the experimentalist faces is which theory is the best for his particular data and how reliable is that theory. Although we have tried to answer these questions, we have not written a book for the complete novice in theoretical chemistry. We would not wish to encourage such a person to use the theories we describe without some knowledge of simple MO theories like Hückel theory. On the other hand we hope to show that SCF theories are not so complicated that they can only be understood and used by the professional theoretician.

Although Hückel theory has had many successes in theoretical organic chemistry we believe that it has now largely outlived its usefulness. The SCF theories which have replaced it are based on firmer theoretical foundations and they are more reliable for the prediction and understanding of quantitative data. We therefore begin the book by emphasizing some of the failures of Hückel theory.

In chapter 2 we describe the SCF theories of π electrons which were developed after 1953. The mathematical developments introduced in this

v

chapter are sufficient to carry the reader through the more recent all-valence-electron theories which are described in chapter 3. These two chapters cover the properties of molecules both in their ground states and in excited electronic states. There has been some difficulty in getting parameters to fit both types of data, which is the reason that most books deal with one or the other. We shall however show that in the more recent theories parameters can be found which are adequate for both.

Chapters 4 and 5 give applications of the theories in the fields of chemical structure and reactivity and in magnetic resonance spectroscopy. These chapters contain some of the subsidiary theory needed to analyse the experimental data. For example, we discuss the theory of the ESR hyperfine constant before describing its interpretation in terms of electronic wave functions. We complete the book with a brief discussion of possible future developments and include two mathematical appendices.

The book lists over 350 references and these have been chosen from a fairly thorough literature search up to the middle of 1970. There is a much wider coverage for the all-valence-electron theories than for the early π-electron SCF theories as we feel that the latter are more widely available in earlier texts.

J. N. M.
A. J. H.

Contents

Symbols and abbreviations

Fundamental constants

e	electronic charge	4.803×10^{-10} e.s.u.
m	electronic mass	9.109×10^{-28} g
c	velocity of light	2.998×10^{10} cm sec^{-1}
$\hbar = h/2\pi$	Planck's constant	1.054×10^{-27} erg sec
k	Boltzmann's constant	1.381×10^{-16} erg degree^{-1}
a_0	Bohr radius	0.5292 Å

Operators

$\mathscr{H}$ The complete electronic Hamiltonian, or a general Hamiltonian

H   a one-electron Hamiltonian not rigorously defined

H^c the core Hamiltonian (the terms in $\mathscr{H}$ which are functions of the coordinates of one electron)

F   the self-consistent-field operator

$\mathbf{J}^2, \mathbf{J}_z, \mathbf{J}, \mathbf{J}^+, \mathbf{J}^-, \mathbf{L}^2, \mathbf{S}^2, \mathbf{I}^2$, etc. angular momentum operators

$$\nabla^2 = \frac{\partial^2}{\partial x^2} + \frac{\partial^2}{\partial y^2} + \frac{\partial^2}{\partial z^2}$$ del squared, which occurs in the kinetic energy operator

Wave functions

φ	an atomic orbital
ψ	a molecular orbital
Ψ	a complete wave function
α, β	spin wave functions

Integrals

$$\int \cdots \mathrm{d}v \qquad \text{integration over space coordinates}$$

$$\int \cdots \mathrm{d}\tau \qquad \text{integration over space and spin coordinates}$$

$$\mathscr{H}_{rs} = \int \Psi_r^* \mathscr{H} \Psi_s \, \mathrm{d}\tau \qquad \text{a matrix element of } \mathscr{H}$$

$$S_{rs} = \int \Psi_r^* \Psi_s \, \mathrm{d}\tau \qquad \text{the overlap integral}$$

$$(\mu\nu \mid \rho\sigma) = \iint \varphi_\mu(1)\varphi_\nu(1)(e^2/r_{12})\varphi_\rho(2)\varphi_\sigma(2) \, \mathrm{d}v_1 \, \mathrm{d}v_2$$

$$\gamma_{\mu\nu} = (\mu^2 \mid \nu^2)$$

Other symbols

A	electron affinity
c	an expansion coefficient
F^n, G^n	Slater-Condon parameters
$\mathbf{H}$	magnetic field
I	ionization potential
k	rate constant
K	equilibrium constant
$P_{\mu\nu}$	bond order
Z_M	nuclear charge
δ_{rs}	Kronecker delta $(= 0, r \neq s; = 1, r = s)$
ζ	Slater orbital exponent
π	polarizability factor
ρ	charge density

Energy conversion table

	erg molecule^{-1}	kcal mole^{-1}	eV	a.u.
erg/molecule	1	$1 \cdot 439 \times 10^{13}$	$6 \cdot 242 \times 10^{11}$	$2 \cdot 294 \times 10^{12}$
kcal/mole	$6 \cdot 947 \times 10^{-14}$	1	$4 \cdot 336 \times 10^{-2}$	$1 \cdot 594 \times 10^{-3}$
eV	$1 \cdot 602 \times 10^{-12}$	$23 \cdot 06$	1	$3 \cdot 675 \times 10^{-2}$
a.u. (Hartree)	$4 \cdot 359 \times 10^{-11}$	$627 \cdot 5$	$27 \cdot 21$	1

Chapter 1

The limitations of Hückel theory and the development of the self-consistent-field (SCF) method

Most chemists are familiar with the Hückel molecular orbital theory of π electrons, and many have followed the recent extensions of the theory to σ electrons. This theory has been very successful in explaining many properties of organic molecules, and has to a large extent superseded the more traditional resonance theory in the language of the experimental chemist. The stimulus of the work of Coulson, Dewar and Longuet-Higgins, and later of Hoffmann and Woodward, has shown that molecular orbital theory in its simplest form leads to quantitative predictions of molecular properties after only a trivial mathematical analysis.

The successes of this simple theory cover the field of molecular geometries, ground-state energies, reactivity and spectroscopic properties. There is no shortage of good books on the subject and we list a few of them in the bibliography.[1-3] Our objective in this chapter is to emphasize the points at which Hückel theory fails as a justification for the development of more comprehensive theories.

The history of calculations on small molecules like H_2 suggests that there is no simple way of obtaining exact solutions of the Schrödinger equation for many-electron atoms and molecules. By exact we mean that the calculated *total* energies have an accuracy comparable with those which can be generally obtained from experiment. Energy differences, deduced by spectroscopic methods, may of course cover a very wide range, depending on the type of spectroscopy used, and 'exact' solutions may show up badly when tested by the criteria of low-frequency spectroscopy.

'Exact' solutions of the electronic Hamiltonian—that is the Hamiltonian based on the Born–Oppenheimer approximation of stationary nuclei—have been obtained for simple molecules[4-6] like H_2, LiH and H_3 and one can expect exact solutions for molecules like H_2O within a few

1

years. However, chemists are generally interested in far larger molecules than these, for which such exact solutions are not possible within the foreseeable future.

Between the extremes of Hückel theory and exact calculations there are many methods of calculating approximate wave functions. Most of these fall within the model called the self-consistent-field molecular-orbital (SCF–MO) method. In this one uses the concept of a molecular orbital, as being the wavefunction of one electron moving in the potential of the nuclei and the average effective potential of the other electrons. Within this model there are various levels of approximation. At the one extreme one has best wave functions of this type, calculated without any empirical parameters, which are called Hartree–Fock wave functions. At the other extreme one has semi-empirical π-electron theories such as that developed by Pople.[7]

To understand the various approximations that have been introduced into the SCF method we must first define the equations on which it is based. The derivation of these equations is not easy to follow without a good background in quantum mechanics, and we have therefore put this derivation in appendix 2 in order to carry the reader through to the more important parts of the book.

In Hückel theory the molecular orbitals ψ are written as a linear combination of atomic orbitals φ (the LCAO approximation)

$$\psi = \sum_v c_v \varphi_v \tag{1.1}$$

and are assumed to be solutions of the equation

$$\mathbf{H}\psi = E\psi \tag{1.2}$$

where $\mathbf{H}$ is a one-electron operator. If expression (1.1) is substituted into (1.2) we obtain

$$\sum_v c_v(\mathbf{H} - E)\varphi_v = 0. \tag{1.3}$$

The coefficients c_v are most easily determined by multiplying (1.3) by one of the atomic orbitals φ_μ and integrating over all the three dimensional space.† This gives the so-called secular equations

$$\sum_v c_v \int \varphi_\mu(\mathbf{H} - E)\varphi_v \, \mathrm{d}v = 0 \tag{1.4}$$

† More precisely one should multiply by the complex conjugate of φ_μ, but as most calculations of this type use real orbitals we shall ignore this feature throughout the book.

If we define the quantities $H_{\mu\nu}$ and $S_{\mu\nu}$ by

$$\int \varphi_\mu \mathbf{H} \varphi_\nu \, dv = H_{\mu\nu}, \tag{1.5}$$

and

$$\int \varphi_\mu E \varphi_\nu \, dv = E \int \varphi_\mu \varphi_\nu \, dv = ES_{\mu\nu}, \tag{1.6}$$

then the equations have the form

$$\sum_\nu c_\nu (H_{\mu\nu} - ES_{\mu\nu}) = 0. \tag{1.7}$$

There is one equation of this type for each atomic orbital φ_μ in the set. To find their solutions the allowed energies are first determined by equating the secular determinant to zero

$$|H_{\mu\nu} - ES_{\mu\nu}| = 0, \tag{1.8}$$

and each energy is then substituted into (1.7) to determine the appropriate set of coefficients.

The essential feature of Hückel-type theories is that the operator $\mathbf{H}$ is not defined by the terms in the complete Hamiltonian of the molecule. Instead it is assumed that the integrals $H_{\mu\nu}$ are meaningful quantities whose values may be determined empirically, that is by fitting theory and experiment.

The theory finds its simplest form in the familiar π-electron theory, in which the energy of repulsion between electrons is ignored, or at best assumed to be a constant which is independent of the detailed distribution of electrons within the molecule. This type of theory is generally called an independent-electron theory, in the sense that the wave function and energy of a molecular orbital do not depend on the number of electrons occupying other molecular orbitals of the molecule. There are modifications of Hückel theory, the Wheland-Mann or the ω-method for example[1], in which some account is taken of electron repulsion, and these can be considered as the simplest form of SCF theories.

The empirical parameters of Hückel theory are usually given the symbols α and β, defined by

$$\alpha_\mu = \int \varphi_\mu \mathbf{H} \varphi_\mu \, dv, \tag{1.9}$$

$$\beta_{\mu\nu} = \int \varphi_\mu \mathbf{H} \varphi_\nu \, dv \tag{1.10}$$

which are called the coulomb and resonance integrals respectively. The overlap integrals $S_{\mu\nu}$ ($\mu \neq \nu$) may be taken as zero, as in Hückel π-electron theory, or they may be calculated from atomic orbital wave functions, as in Hoffmann's extended-Hückel theory of σ electrons[8].

At this point we shall list a few of the unsatisfactory features of Hückel theory which have provided the stimulation for the development of the more sophisticated theories which are to be described in this book.

1. In general, values of the Hückel parameters α and β are determined empirically by fitting theoretical calculations to observed data. However, the optimum value of the parameters depends on the nature of the experimental data that is under consideration. For example, the first ionization potentials of aromatic hydrocarbons correlate closely with the energy of the highest occupied Hückel π orbital if this is calculated with $\beta = -4\cdot0$ eV. For the same compounds, the energy of the first strongly allowed electronic transition (giving the so-called ^{1}La or p-band) is given by the difference in energy between the lowest vacant and highest occupied molecular orbitals calculated with $\beta = -2\cdot4$ eV. Finally, if we examine the correlation between the Hückel delocalization energy (that is the total energy less a value of 2β for each formal double bond) and the observed resonance energy (the difference between the total energy and that calculated on the basis of additive bond energies), then from the correlation line we deduce $\beta = -0\cdot7$ eV.†

From the very large difference between these three values for β one would conclude that the parameter represents a different combination of one-electron energies (nuclear attraction and kinetic energy) and two-electron repulsion energies, for each different type of data. This is what we find to be the case when we examine the relevant energy expressions of scf theory.

2. Hückel theory is more successful for non-polar molecules, such as hydrocarbons, than for molecules with relatively polar bonds. It is more successful for alternant aromatic hydrocarbons (e.g. naphthalene), which in their ground states have a uniform distribution of π charge over the carbon atoms, than for non-alternant hydrocarbons (e.g. azulene), which have a non-uniform distribution of π charge. Thus the dipole moment of azulene, calculated from Hückel theory, is about seven times larger than the experimental value. This discrepancy arises from the neglect of electron repulsion which is inherent in the assumption that α is independent of the charge on the atom.

† These parameters are all as deduced by Salem.[3]

3. In calculations on hetero-atomic molecules the number of empirical parameters (α_X, β_{XY}) needed to carry out a calculation may be very large whereas the amount of experimental data from which to determine these parameters may be relatively small. This problem is present to some extent in the semi-empirical SCF theories, but as these make use of a considerable amount of atomic spectral data it is not usually so severe. If in Hückel theory one attempts to take the coulomb integrals directly from atomic spectral data then one usually ends up with unreasonably large bond polarities, again because of the neglect of electron repulsion. Moreover, it is often necessary to allow for the fact that α_X for an X—Y molecular fragment depends on the nature of atom Y.

4. There are some molecular properties, notably the energies and intensities of some electronic absorption bands, which cannot be fitted at all by Hückel theory, or by any theory based on a one-electron Hamiltonian. For example, amongst the π molecular orbitals of benzene both the highest occupied and lowest unoccupied molecular orbitals are doubly degenerate. If the excitation energy is equated to the difference in the orbital energies, which would follow if electron repulsion were ignored, then one would predict a four-fold-degenerate excited state and one strong band in the absorption spectrum. It is known, however, that such an electronic excitation gives three excited states, which give two weak (symmetry-forbidden) bands in the absorption spectrum and one strong band.

There are several books[9-12] that describe in detail the theory of the electronic spectra of organic molecules, and we shall not cover this topic to any large extent in this book. We wish to emphasize however that the failure of Hückel theory to explain spectroscopic data played a large part in the initial development of the self-consistent-field theories we shall describe in this book.

The SCF molecular orbitals are defined by formally similar equations to (1.7) and (1.8). They are taken to be eigenfunctions of an operator $\mathbf{F}$

$$\mathbf{F}\psi = E\psi \qquad (1.11)$$

and if the LCAO expansion (1.1) is adopted, then the coefficients and energies are determined by the equations

$$\sum_v c_v(F_{\mu v} - ES_{\mu v}) = 0, \qquad (1.12)$$

$$|F_{\mu v} - ES_{\mu v}| = 0, \qquad (1.13)$$

where

$$F_{\mu v} = \int \varphi_\mu \mathbf{F} \varphi_v \, dv. \qquad (1.14)$$

However, unlike the Hückel operator, $\mathbf{F}$ is well defined by the elements of the full Hamiltonian. Its matrix elements, $F_{\mu\nu}$, are given by expressions first derived by Lennard-Jones,[13] Hall[14] and Roothaan.[15]

$$F_{\mu\nu} = H_{\mu\nu}{}^c + \sum_\rho \sum_\sigma P_{\rho\sigma}[(\mu\nu \mid \rho\sigma) - \tfrac{1}{2}(\mu\rho \mid \nu\sigma)] \tag{1.15}$$

H^c is called the core Hamiltonian for an electron. It consists of the kinetic energy operator for an electron and the potential energy between an electron and all atomic cores of the molecule.

$$\mathbf{H}^c = \frac{-\hbar^2}{2m} \mathbf{\nabla}^2 + \sum_A V_A, \tag{1.16}$$

where

$$\mathbf{\nabla}^2 = \partial^2/\partial x^2 + \partial^2/\partial y^2 + \partial^2/\partial z^2. \tag{1.17}$$

If all electrons are being specifically included in the calculation then V_A is the nuclear-electron potential energy equal to $-Z_A e^2/r_A$ (Z_A being the nuclear charge). In a π-electron model V_A would be the potential energy of the nuclei together with the repulsion of the σ electrons. The complete electronic Hamiltonian, $\mathscr{H}$, is made up of the core terms and the potential energy of repulsion of the electrons

$$\mathscr{H} = \sum_i \mathbf{H}^c(i) + \sum_{i<j} e^2/r_{ij} \tag{1.18}$$

It is usual in quantum-mechanical calculations on molecules to work in a system of units called atomic units in which the charge and mass of the electron and $\hbar$ are all taken as unity. In these units the unit of length is the Bohr radius $a_0 = 0\cdot5292$ Å and the unit of energy (called the Hartree) is $27\cdot21$ eV. In these units the complete Hamiltonian has the form

$$\mathscr{H} = -\sum_i \tfrac{1}{2}\mathbf{\nabla}_i{}^2 + \sum_A V_A + \sum_{i<j} r_{ij}{}^{-1} \tag{1.19}$$

The remaining terms in (1.15) give the effect of the electron interaction. We use the definition

$$(\mu\nu \mid \rho\sigma) = \iint \varphi_\mu(1)\varphi_\nu(1)(e^2/r_{12})\varphi_\rho(2)\varphi_\sigma(2) \, dv_1 \, dv_2 \tag{1.20}$$

which is to be interpreted physically as the repulsion between an electron distributed in space according to the function $\varphi_\mu \varphi_\nu$ (1) and a second electron having the distribution $\varphi_\rho \varphi_\sigma$ (2).

The final term to be defined in (1.15) is the bond order $P_{\rho\sigma}$ which is written

$$P_{\rho\sigma} = 2 \sum_k c_{k\rho} c_{k\sigma}, \tag{1.21}$$

the summation extending over all *occupied* molecular orbitals ψ_k. In π-electron theory $P_{\rho\sigma}$ has special significance when ρ and σ are atoms joined together because it has been found to be linearly related to the length of the ρ—σ bond.

Expression (1.15) only applies to closed-shell electron configurations, that is when all occupied molecular orbitals contain two electrons. It therefore applies to the ground states of most molecules. For radicals, or most excited electronic states slightly different equations are required.[16]

Once the elements $F_{\mu\nu}$ and the overlap integrals $S_{\mu\nu}$ are known the SCF orbitals are obtained with the same ease as the Hückel orbitals. There are however two major difficulties. In the first place one can see from (1.15) that $F_{\mu\nu}$ depends on the bond orders, and these from their definition (1.21) can only be calculated when the orbitals, that is the solutions of (1.12) and (1.13), are known. The equations have therefore to be solved iteratively. A rough estimate is made of the coefficients $c_{k\rho}$ (usually these are taken as the coefficients obtained from a Hückel calculation), which then allows one to make an estimate of the bond orders and $F_{\mu\nu}$ integrals. The secular equations are then solved to give improved values of the coefficients. The cycle of the calculation can be repeated until the coefficients obtained by solving the secular equations are the same as those used to construct $F_{\mu\nu}$: that is the input and output coefficients are self-consistent. In practice if a reasonable first estimate is made, the SCF cycle is convergent, but divergent situations can be encountered.

The second difficulty in solving the SCF equations lies in the evaluation of the integrals involved in $F_{\mu\nu}$, particularly those two-electron integrals (1.20) in which the four orbitals are all on different atomic orbitals.

Except for one-electron atoms, atomic orbitals are not simple functions of the distance between the electron and the nucleus. Accurate atomic orbitals are either expressed in tabular form, $\varphi(\mathbf{r})$ tabulated as a function of r, the distance of the electron from the nucleus, or as a linear combination of simple algebraic functions. It is known from the asymptotic form of the solutions of the Schrödinger equation that at large r, $\varphi(\mathbf{r})$ varies as $\exp(-kr)$, thus the most convenient functions from which to build up accurate atomic orbitals are the so-called Slater orbitals

$$\chi_{nlm}(k, \mathbf{r}) = Nr^{n-1}\exp(-kr)Y_{lm}(\theta, \phi) \tag{1.22}$$

n is an integer, which corresponds to the principal quantum number. The $Y_{lm}(\theta, \phi)$ are the spherical harmonic functions which describe the angular variation of the orbital, and are labelled by the quantum numbers l and m.

If Slater orbitals are used in the LCAO expansion (1.1), that is the φ_ν are taken as χ or as some linear combination of such functions, then the three and four-centre integrals are difficult to evaluate even on a large computer. They must either be obtained by numerical integration or by a series expansion with relatively slow convergence. One of the reasons that approximate SCF schemes have been developed is to obviate the necessity for calculating such integrals. This development will be described in the coming chapters.

An alternative set of functions that have been used to build up atomic orbitals are gaussian functions which have a radial dependence $\exp(-kr^2)$. These do not have the correct asymptotic limit, but if enough of these functions are taken in a linear expansion then it is possible to get a reasonable representation of an atomic orbital. As a rough guide one can say that if a Slater orbital is replaced by a sum of 3 to 5 gaussians then the results of the SCF calculations obtained with the Slater or gaussian basis will be similar for most properties of chemical interest. Although one needs to use more gaussian orbitals than Slater orbitals to obtain the same accuracy in an SCF calculation, and therefore more two-electron integrals like (1.20) occur in the calculation, this is more than offset by the relative ease with which the integrals can be evaluated. The reason for this is that the product of two gaussian orbitals on different centres, $\varphi_\mu \varphi_\nu$ is equal to a third gaussian centred somewhere between the two[17]. Because of this all three and four-centre integrals reduce to two-centre integrals which are relatively easy to calculate.

There are several different ways in which gaussian orbitals have been used in SCF calculations. One approach has been to use gaussian orbitals just to calculate the two-electron integrals over Slater orbitals. Another approach has been to use them directly as the LCAO expansion functions in (1.1). In the latter method one may choose to use only spherical gaussians, and then angular functions like p and d orbitals are built up by taking combinations of off-nuclear gaussians. Alternatively one can use the so-called cartesian gaussians like $x \exp(-kr^2)$ which have an angular dependence like spherical-harmonic functions ($x = r \cos \theta$, has an angular variation like a p orbital). It is not the main task of this book to give a critical account of the large number of calculations made with gaussian orbitals, even on quite large organic molecules like benzene[18] and pyridine.[19]. At the present time better calculations can be made on medium-sized molecules using a gaussian basis than a Slater basis, however the balance between the two methods can change with an advance in computer design or the techniques of numerical analysis.

The number of molecular orbitals, obtained in an LCAO calculation is equal to the number of atomic orbitals in the expansion. If the molecule has $2n$ electrons then clearly the absolute minimum number of atomic orbitals needed is n. In practice the minimum number will be greater than n as it will consist at least of all the atomic orbitals that are occupied by electrons in the ground states of the separate atoms. For hydrocarbons this means one orbital for each hydrogen ($1s$) and five for each carbon atom ($1s$, $2s$, $2px$, $2py$, $2pz$). In the absence of symmetry in the molecule a non-empirical SCF calculation with n atomic orbitals in the expansion will require the calculation of n^4 two-electron integrals and the solution of the n secular equations. Either of these can be the limiting factor which determines the maximum size of a molecule that is amenable even to a minimum-basis non-empirical calculation.

However, if empirical methods are to be used, the size of the molecule that can be examined may be increased considerably. In the first place some of the n^4 two-electron integrals may be neglected or be given empirical values. In the second place it may be possible to consider in detail only the electrons in the outer shells of the atoms, using the argument that inner-shell electrons (e.g. the $1s$ electrons of carbon) are little affected by bond formation. Thus the number of atomic orbitals that may be considered in an empirical calculation may be significantly less than the number required for a non-empirical calculation.

There is one other justification for the empirical calculation, and this is that we know that even an exact Hartree–Fock calculation would not give accurate values of all the quantities we are interested in because it is not a solution of the Schrödinger equation. An example of this is the prediction[20] by the Hartree–Fock method that F_2 is unstable. However, an empirical calculation may be parameterized in such a way that for a family of compounds it does give reliable values of the experimental quantities of chemical interest. We will return to this point later in the book.

In the following chapters we deal firstly with the SCF developments of π-electron theory and then with the extension to σ electrons. Applications will be given to a wide range of chemical properties, both for ground and excited states, and to the interpretation of magnetic resonance properties. We will conclude by giving our view of the likely development of the subject.

References

1. A. Streitweiser, Jnr., *Molecular Orbital Theory for Organic Chemists*, Wiley, New York, 1961.

2. E. Heilbronner and H. Bock, *Das HMO-modell und Seine Anwendung*, Verlag Chemie, 1968.
3. L. Salem, *Molecular Orbital Theory of Conjugated Systems*, Benjamin, New York, 1966.
4. W. Kølos and C. C. J. Roothaan, *Rev. Mod. Phys.*, **32**, 219 (1960).
5. S. F. Boys and N. C. Handy, *Proc. Roy. Soc. (London)*, **A311**, 309 (1969).
6. H. Conroy and B. L. Bruner, *J. Chem. Phys.*, **47**, 921 (1967).
7. J. A. Pople, *Trans. Faraday Soc.*, **49**, 1375 (1953).
8. R. Hoffmann, *J. Chem. Phys.*, **39**, 1397 (1963).
9. J. N. Murrell, *The Theory of the Electronic Spectra of Organic Molecules*, Chapman and Hall, London, 1971.
10. C. N. R. Rao, *Ultraviolet and Visible Spectroscopy*, Butterworth, London, 1961.
11. H. H. Jaffé and M. Orchin, *Theory and applications of Ultraviolet Spectroscopy*, Wiley, New York, 1962.
12. C. Sandorfy, *Electronic Spectra and Quantum Chemistry*, Prentice-Hall, New Jersey, 1964.
13. J. E. Lennard-Jones, *Proc. Roy. Soc. (London)*, **A198**, 1, 14 (1949).
14. G. G. Hall, *Proc. Roy. Soc. (London)*, **A205**, 541 (1951).
15. C. C. J. Roothaan, *Rev. Mod. Phys.*, **23**, 69 (1951).
16. C. C. J. Roothaan, *Rev. Mod. Phys.*, **32**, 179 (1960).
17. S. F. Boys, *Proc. Roy. Soc. (London)*, **A200**, 542 (1950).
18. R. J. Buenker, J. L. Whitten and J. D. Petke, *J. Chem. Phys.*, **49**, 2261 (1968).
19. E. Clementi, *J. Chem. Phys.*, **46**, 4731 (1967).
20. A. C. Wahl, *J. Chem. Phys.*, **41**, 2600 (1964).

Chapter 2

The Pariser–Parr–Pople π-electron theory and its development

2.1 The Zero–Differential–Overlap models

A wave function φ is said to be normalized if

$$S_{\mu\mu} = \int \varphi_\mu \varphi_\mu \, \mathrm{d}v = 1. \tag{2.1}$$

Taking $\varphi_\mu{}^2$ as having the physical significance of a probability density for an electron with wave function φ_μ this condition fulfills the requirement that there is unit probability of finding the electron somewhere in space. In Hückel π-electron theory the overlap integral $S_{\mu\nu}$ is assumed to be zero even between atomic orbitals on neighbouring centres. This condition together with (2.1) can be combined in the expression

$$S_{\mu\nu} = \delta_{\mu\nu} \tag{2.2}$$

where $\delta_{\mu\nu}$ (the Kronecker delta) is unity if $\mu = \nu$ and zero otherwise. This approximation converts the secular determinant (1.8) into

$$|H_{\mu\nu} - E \, \delta_{\mu\nu}| = 0 \tag{2.3}$$

and this is certainly a little easier to solve by hand than (1.8) because there are fewer terms in the expansion of the determinant. However, the resulting simplification is not very significant with modern computing facilities because one can see by matrix multiplication that equation (1.8) can be replaced by

$$|H_{\mu\nu}' - E \, \delta_{\mu\nu}| = 0 \tag{2.4}$$

if the matrix $\mathbf{H}'$ is defined by

$$\mathbf{H}' = \mathbf{S}^{-1}\mathbf{H} \quad \text{or} \quad \mathbf{S}^{-\frac{1}{2}}\mathbf{H}\mathbf{S}^{-\frac{1}{2}}. \tag{2.5}$$

11

The importance of the zero-overlap approximation in SCF theory is that it enables one to reduce the number of two-electron integrals that need be considered. If we return to expression (1.20) we see that unless there is some region of space in which φ_μ and φ_ν are simultaneously nonzero, and also some region in which φ_ρ and φ_σ are simultaneously nonzero, then $(\mu\nu \mid \rho\sigma) = 0$. A complete zero overlap of φ_μ and φ_ν is sufficient to make this integral zero and to make $S_{\mu\nu} = 0$. Of course special circumstances may make either of these integrals zero even if there is some region of overlap of φ_μ and φ_ν. For example, in the case of s and p orbitals on the same atom, $S_{\mu\nu} = 0$, and yet $(\mu\nu \mid \mu\nu) \neq 0$ because the two orbitals do occupy the same region of space. Nevertheless, we have sufficient grounds for saying that if we are going to make the approximation that $S_{\mu\nu} = 0$ or $S_{\rho\sigma} = 0$ for orbitals on different atoms then to be consistent we should also make $(\mu\nu \mid \rho\sigma) = 0$.

A consistent zero-differential-overlap (ZDO) approximation was first made by Pariser and Parr[1,2] in π-molecular-orbital theory, and by Pople[3] who gave the resulting expressions for the SCF operator. We will describe this work before passing to the more general theories that have since been developed.

In the ZDO approximation if φ_μ and φ_ν are different orbitals, then only one electron repulsion integral arises from the summation over ρ and σ in (1.15), and that occurs when $\rho = \mu$ and $\sigma = \nu$, to give the integral

$$(\mu\mu \mid \nu\nu) \equiv \gamma_{\mu\nu}. \tag{2.6}$$

This integral is usually represented by the symbol $\gamma_{\mu\nu}$. The off-diagonal elements of the F matrix, (1.15),

$$F_{\mu\nu} = H_{\mu\nu}{}^c + \sum_\rho \sum_\sigma P_{\rho\sigma}[(\mu\nu \mid \rho\sigma) - \tfrac{1}{2}(\mu\rho \mid \nu\sigma)] \tag{2.7}$$

therefore have the form

$$F_{\mu\nu} = H_{\mu\nu}{}^c - \tfrac{1}{2}P_{\mu\nu}\gamma_{\mu\nu}. \tag{2.8}$$

The core matrix element is, from (1.16)

$$H_{\mu\nu}{}^c = \int \varphi_\mu \left(-\tfrac{1}{2}\nabla^2 + V_M + V_N + \sum_A{}'' V_A \right) \varphi_\nu \, \mathrm{d}v. \tag{2.9}$$

Here we have separated the core potentials V_M and V_N of atoms M and N, (which are the nuclear centres of φ_μ and φ_ν respectively), from the remainder, $A \neq M, N$. In a strictly zero-overlap model it would seem logical to take this integral as zero also. However if this is done one loses the essential bond forming term in molecular orbital theory, because $H_{\mu\nu}{}^c$ represents the energy of attraction of the overlap cloud (between the orbitals φ_μ and φ_ν) for the positively charged core. Therefore one must assume that there is sufficient overlap of φ_μ and φ_ν, at least for orbitals on neighbouring atoms, to give a nonzero integral. In order to have a parameter which can be carried over as characteristic of the M—N bond, we must assume that the potentials of distant cores $V_A(A \neq M, N)$ make a negligible contribution to the integral (2.9). One can then define a 'resonance integral' β as in Hückel theory by

$$\beta_{\mu\nu} = \int \varphi_\mu(-\tfrac{1}{2}\nabla^2 + V_M + V_N)\varphi_\nu \, dv \qquad (2.10)$$

and the off-diagonal matrix elements of **F** become, from (2.8)

$$F_{\mu\nu} = \beta_{\mu\nu} - \tfrac{1}{2}P_{\mu\nu}\gamma_{\mu\nu} \qquad (2.11)$$

For the diagonal elements of the **F** matrix ($\mu = \nu$) the ZDO approximation leads to

$$F_{\mu\mu} = H_{\mu\mu}{}^c + \sum_\rho P_{\rho\rho}(\mu\mu \mid \rho\rho) - \tfrac{1}{2}P_{\mu\mu}(\mu\mu \mid \mu\mu) \qquad (2.12)$$

The core Hamiltonian integral $H_{\mu\mu}{}^c$ can be treated in the same way as for the off-diagonal elements. We write

$$H_{\mu\mu}{}^c = \int \varphi_\mu\left(-\tfrac{1}{2}\nabla^2 + V_M + \sum_A{}' V_A\right)\varphi_\mu \, dv, \qquad (2.13)$$

$$= U_{\mu\mu} + \sum_A{}' \int \varphi_\mu V_A \varphi_\mu \, dv, \qquad (2.14)$$

where $A \neq M$, and

$$U_{\mu\mu} = \int \varphi_\mu(-\tfrac{1}{2}\nabla^2 + V_M)\varphi_\mu \, dv, \qquad (2.15)$$

can be taken as the energy of the orbital φ_μ for the appropriate valence state of the isolated atom M. The valence state of an atom is a function of the hybridization assumed for the atom—it is not, in general, a real spectroscopic state of the atom, but its energy can be calculated in terms of spectroscopic energies.[4]

If atoms A and M are far apart then the integral

$$\int \varphi_\mu V_A \varphi_\mu \, dv \equiv V_{A,\mu\mu} \qquad (2.16)$$

will be approximately equal to $-Z_A e^2 R_{AM}{}^{-1}$ where $-Z_A e$ is the net charge of the atomic core of atom A. A similar dependence on R_{AM} is shown by the two-centre electron repulsion integrals: at large separations of the two atoms

$$\gamma_{\mu\rho} = (\mu\mu \mid \rho\rho) = e^2 R_{AM}{}^{-1} \qquad (2.17)$$

where orbital φ_ρ is on atom A. We can therefore write

$$V_{A,\mu\mu} = -f(R)Z_A\gamma_{\mu\rho} \tag{2.18}$$

where $f(R)$ is called the penetration function which allows for the deviation of $V_{A,\mu\mu}$ and $\gamma_{\mu\rho}$ from R_{AM}^{-1} at small R_{AM}. In fact for all the ZDO theories it has been found that satisfactory results are obtained by having $f(R) = 1$, that is, when the penetration effect is ignored.

Substituting (2.14) and (2.18) into (2.12) gives

$$F_{\mu\mu} = U_{\mu\mu} + \tfrac{1}{2}P_{\mu\mu}\gamma_{\mu\mu} + \sum_{\rho \neq \mu}(P_{\rho\rho} - Z_A)\gamma_{\mu\rho}. \tag{2.19}$$

Expressions (2.11) and (2.19) define the elements of the Pople SCF equations for π electrons.

Although the ZDO approximation may seem to be rather severe (e.g. the overlap integral between π orbitals on two carbon atoms separated by $1\cdot4$ Å has the value $0\cdot25$, and this is neglected), the Pople SCF method has been applied with remarkable success to a wide range of chemical problems. Part of the success lies in the fact that the treatment is semi-empirical, so that any errors introduced by use of the ZDO approximation can be partially compensated by a judicious choice of parameters. However, some difficulty has been found in obtaining one set of parameters which can satisfactorily explain both spectroscopic and ground-state properties. This will become apparent in the following section.

2.2   Choice of SCF π-electron parameters

In the ZDO π-electron theories the atomic orbitals are usually taken to be Slater functions, and all the two-electron integrals $\gamma_{\mu\nu}$ can be calculated quite easily. However, when Pariser and Parr used these non-empirical integrals to calculate the excitation energies of unsaturated hydrocarbons the results were discouragingly bad.[1] In a subsequent calculation[2] they gave some of these integrals semi-empirical values and obtained much better agreement with experiment. One of the modifications introduced by them was to evaluate the one-centre Coulomb integral from the following expression first given by Pariser[5]

$$\gamma_{\mu\mu} = I - A \tag{2.20}$$

Pariser reasoned that the difference between the ionization potential and electron affinity of a carbon atom in the $\pi(sp^2)^3$ valence state should be equal to the repulsion of two electrons in the π orbital, $\gamma_{\mu\mu}$. This follows because the electron affinity is defined as the ionization potential of the negative ion.

The value of $\gamma_{\mu\mu}$ calculated from a Slater orbital with exponent 1·59 is 16·93 eV,[6] but estimates of $I - A$ are around 10 eV. Following the principle of Moffit's 'atoms in molecules' method,[7] which is that the atomic parts of molecular energy calculations should agree with atomic experimental data, Pariser and Parr proposed that $\gamma_{\mu\mu}$ should be determined empirically from expression (2.20).

There are several reasons why the calculated $\gamma_{\mu\mu}$ should not agree with the experimental $I - A$. Firstly, Slater-orbitals are rather too compact: SCF atomic orbitals, which are rather more difficult to use, because they are either in numerical form or are a sum of Slater orbitals, give $\gamma_{\mu\mu} = 15·71$ eV.[8] Secondly, the orbitals of an atom do not remain constant during the ionization process; one would certainly expect the orbitals of C^- to be more diffuse than those of C^+ because of the increased electron repulsion. Thirdly, there are contributions to the energy which depend on the instantaneous relative positions of electrons rather than on their average relative positions. These correlation energies are not given correctly by the Hartree–Fock model. By choosing an empirical value of $\gamma_{\mu\mu}$ a correction can be made for the neglect of these terms in a molecular calculation. A recent Hartree–Fock calculation[9] on the reaction $2C \rightarrow C^+ + C^-$, the energy of which is $I - A$, shows that the Hartree–Fock orbital which should be used to calculate $\gamma_{\mu\mu}$ is that appropriate to a valence state of C^-. This orbital gives $\gamma_{\mu\mu} = 12·72$ eV which is close to the semi-empirical value (see also refs. 10 and 11).

The calculated value of γ_{12} for π orbitals on two neighbouring carbon atoms separated by 1·39 Å is 9·03 eV. This now looks too high in comparison with the chosen empirical value of $\gamma_{\mu\mu}$. Various methods have been proposed for scaling the calculated $\gamma_{\mu\nu}$, or for choosing them empirically. Most methods recognize one or both of the following boundary conditions for $\gamma_{\mu\nu}$:

$$\underset{R \to 0}{\text{Limit}} \, \gamma_{\mu\nu} = \gamma_{\mu\mu} \qquad (2.21)$$

and

$$\underset{R \to \infty}{\text{Limit}} \, \gamma_{\mu\nu} = e^2 R^{-1} \qquad (2.22)$$

The following have been the most commonly used empirical relationships.

1. In the uniformly charged sphere model of Pariser and Parr,[2,12] $\gamma_{\mu\nu}$ is calculated from the following expressions: for $R > 2·8$ Å

$$\gamma_{\mu\nu} = \frac{e^2}{2R} \left\{ \left[1 + \left(\frac{d_\mu - d_\nu}{2R} \right)^2 \right]^{-\frac{1}{2}} + \left[1 + \left(\frac{d_\mu + d_\nu}{2R} \right)^2 \right]^{-\frac{1}{2}} \right\} \qquad (2.23)$$

which is the classical electrostatic repulsion between π-electron densities which are each represented by two tangentially touching charged spheres of diameter

$$d(\text{Å}) = 9 \cdot 194/\zeta \qquad (2.24)$$

ζ being the exponent of the Slater π atomic orbital. For $R < 2 \cdot 8$ Å an extrapolation to the one-centre limit is made through the expression

$$\gamma_{\mu\nu} = \tfrac{1}{2}(\gamma_{\mu\mu} + \gamma_{\nu\nu}) - (AR + BR^2) \qquad (2.25)$$

where the constants A and B are obtained from the value of $\gamma_{\mu\nu}$ given by expression (2.23) at $R = 2 \cdot 8$ Å and $3 \cdot 7$ Å. There is a small discontinuity $\sim 0 \cdot 14$ eV at $R = 2 \cdot 8$ Å arising from the use of two different expressions (see figure 2.1). These are historically important expressions but are now little used.

2. The Mataga–Nishimoto relationship is[13]

$$\gamma_{\mu\nu} = \frac{e^2}{R + a_{\mu\nu}}, \qquad (2.26)$$

where

$$a_{\mu\nu} = \frac{2e^2}{(\gamma_{\mu\mu} + \gamma_{\nu\nu})}. \qquad (2.27)$$

This has been the most popular approximation in SCF treatments of molecular spectra.

3. The Ohno–Klopman approximation has the form[14,15]

$$\gamma_{\mu\nu} = \frac{e^2}{\left[R^2 + \dfrac{e^2}{4}\left(\dfrac{1}{\gamma_{\mu\mu}} + \dfrac{1}{\gamma_{\nu\nu}} \right)^2 \right]^{\frac{1}{2}}}. \qquad (2.28)$$

Values of $\gamma_{\mu\nu}$ obtained from these three relationships, together with the Slater values are shown in figure 2.1. It is seen that $\gamma_{\mu\nu}$ obtained from the Mataga–Nishimoto relationship is much smaller than those obtained from the other three and hence reflects greater electron correlation.[16]

In the work of Pariser and Parr the resonance integral $\beta_{\mu\nu}$ was taken to be zero for all orbitals except those on neighbouring atoms, and this nearest neighbour value was obtained by fitting the spectra of ethylene and benzene and using an exponential extrapolation for intermediate bond lengths.[2] In addition several other relationships have been proposed for evaluating $\beta_{\mu\nu}$, all of which take account of the dependence of $\beta_{\mu\nu}$ on bond length as required by (2.9): the following have been the most widely used.

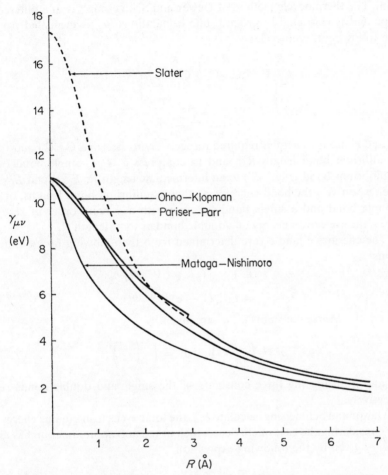

FIGURE 2.1 Theoretical and semi-empirical evaluation of $\gamma_{\mu\nu}$ as a function of R

1. The Mulliken approximation is[17]

$$\beta_{\mu\nu} = kS_{\mu\nu} \tag{2.29}$$

where k is a parameter obtained by fitting theory to experiment for each type of atom pair. The overlap integral $S_{\mu\nu}$ can be easily calculated.

2. The Wolfsberg–Helmholz approximation is[18]

$$\beta_{\mu\nu} = k(U_{\mu\mu} + U_{\nu\nu})S_{\mu\nu} \tag{2.30}$$

where U is defined by (2.15). As in the Mulliken approximation k has to be determined empirically. It usually has a value between 0·5 and 1.

3. The thermocycle method of Dewar and Schmeising[19] is of a different type and is restricted to ground-state calculations. $\beta_{\mu\nu}$ is evaluated from the following thermocycle:

$$\begin{array}{ccccccc} R' & c' & R & E_{\pi b} & R & -c'' & R'' \\ \text{C--C} & \rightarrow & \text{C--C} & \rightarrow & \text{C=C} & \rightarrow & \text{C=C} \end{array} \qquad (2.31)$$

$$E_{\text{C=C}} - E_{\text{C--C}}$$

c'' and c' are the energies required respectively to stretch a C=C bond (of equilibrium bond length R''), and to compress a C—C single bond (of equilibrium bond length R') to an intermediate length R. $E_{\text{C=C}}$ and $E_{\text{C--C}}$ are respectively the bond energies, at their equilibrium bond lengths, of a double bond and a single bond (i.e. one with a π bond order of zero). $E_{\pi b}$ is the π-electron energy of a double bond at some length R.

The energies c' and c'' are determined from the following Morse functions:

$$c' = E_{\text{C--C}}(1 - e^{-a'(R'-R)})$$
$$c'' = E_{\text{C=C}}(1 - e^{-a''(R''-R)}), \qquad (2.32)$$

where the Morse constants a', a'' are given by

$$a' = \left(\frac{k'}{2E_{\text{C--C}}}\right)^{\frac{1}{2}}, \quad a'' = \left(\frac{k''}{2E_{\text{C=C}}}\right)^{\frac{1}{2}}, \qquad (2.33)$$

k' and k'' being the force constants of the single and double bonds respectively.

Dewar and Schmeising calculate $E_{\pi b}$, the total π-electron binding energy, by considering the case of ethylene which in the ZDO approximation has an energy given by the following expression

$$E_{\pi b} = 2\beta + \tfrac{1}{2}(\gamma_{11} - \gamma_{12}). \qquad (2.34)$$

Thus if $E_{\pi b}$ is evaluated from the thermocycle (2.31) expression (2.34) provides an estimate of β.

Those terms in the thermocycle which refer to the double bond, namely c'', k'' and $E_{\text{C=C}}$, present no problems because experimental data are readily available. However, the terms c', k' and $E_{\text{C--C}}$ refer to the hypothetical pure C—C σ bond between sp^2 carbon atoms for which of course no experimental properties are known. The procedure adopted to determine these quantities was firstly to estimate R' from the well-known bond length–bond order relationship

$$R_{\mu\nu}(\text{Å}) = A - BP_{\mu\nu}. \qquad (2.35)$$

Various values for A and B have, in the past, been proposed, none of which have gained general acceptance. Secondly, E_{C-C} and a' were estimated from empirical relationships between bond energy and bond length, and between force constant and bond length. The thermocycle data as now used by Dewar and coworkers[20,21] for the calculation of β are summarized in table 2.1

TABLE 2.1   Thermocycle data used to calculate $\beta^{20,21}$

Bond	$E_{C=C}$	$E_{C=C}$(eV)	R''	R'(Å)	a''	a'(Å^{-1})	A	B(Å)
C—C	5·5600	3·9409	1·338	1·512	2·3177	2·0022	1·512	0·174
C—N	5·1766	3·3463	1·270	1·448	2·5161	1·9209	1·448	0·178
C—O	7·1011	3·9987	1·230	1·395	2·1787	1·7870	1·395	0·165
N—N	4·0926	2·3017	1·240	1·417	3·1787	2·5290	1·417	0·177

4. The Linderberg approximation which has the form[22]

$$\beta_{\mu\nu} = \left(\frac{\hbar^2}{m}\right) R^{-1} \left(\frac{dS_{\mu\nu}}{dR}\right) \qquad (2.36)$$

stands out from the others in having no empirical parameters. This expression was derived from the condition that transition probabilities between molecular orbitals calculated from the dipole-velocity or dipole-moment matrix elements should be equal. If molecular orbitals ψ_i and ψ_j are eigenfunctions of $\mathbf{F}$, then one can derive the following identity

$$(E_i - E_j)\langle\psi_i| \, x \, |\psi_j\rangle = \langle\psi_i| \, \mathbf{F}x - x\mathbf{F} \, |\psi_j\rangle = -\frac{\hbar^2}{m} \langle\psi_i| \, \partial/\partial x \, |\psi_j\rangle. \qquad (2.37)$$

This expression follows from the fact that the only term in $\mathbf{F}$ which does not commute with x is the part of the kinetic energy $-(\hbar^2/2m)\,(\partial^2/\partial x^2)$. When the right hand side of (2·37) is expanded in terms of atomic orbitals one obtains an expression involving derivatives of the overlap integrals. The left hand scale of (2·37) is the product of an energy and a dipole moment. In so far as the energy depends on β one can see qualitatively the origins of expression (2·36). However, as $E_i - E_j$ depends on energy terms other than β, for example the electron repulsion integrals, there is no exact relationship between β and (dS/dR). We do not give the details of Linderberg's derivation of (2·36) because it appears not to be rigorous, however the relationship is quoted here as it has been used successfully in calculations to be described later.

5. The Nishimoto–Forster approximation directly relates $\beta_{\mu\nu}$ to the bond order $P_{\mu\nu}$ in the following manner[23]

$$\beta_{\mu\nu} = A + BP_{\mu\nu}. \qquad (2.38)$$

It can therefore only be used for nearest-neighbour interactions. The parameters A and B were obtained by fitting calculated and spectroscopic energies.

We do not wish to commit ourselves on the relative merits of these expressions since their success depends on the type of data under consideration, the choice of empirical parameters and finally on the method of choosing $\gamma_{\mu\nu}$.

For conjugated hydrocarbons the excitation energies depend only on the values of $\beta_{\mu\nu}$ and $\gamma_{\mu\nu}$. For heteroatomic conjugated systems the relative values of $U_{\mu\mu}$ also need to be known. It is usual to make use of Koopmans' theorem[24] which states that an orbital energy may be equated to minus the ionization potential for removal of an electron from that orbital. Providing that the correct valence states of the atom and ion are used we can therefore take

$$U_{\mu\mu} = -I_\mu. \qquad (2.39)$$

Valence-state ionization potentials are usually taken from the data of either Pilcher and Skinner[25] or the more popular set of Hinze and Jaffé.[26] Values of $U_{\mu\mu}$ and $\gamma_{\mu\mu}$ calculated from both sets of data are given in table 2.2 for carbon, nitrogen and oxygen.

TABLE 2.2   Atomic parameters deduced from valence state energies

Atom	Pilcher and Skinner[25]		Hinze and Jaffé[26]	
	$U_{\mu\mu}$(eV)	$\gamma_{\mu\mu}$(eV)	$U_{\mu\mu}$(eV)	$\gamma_{\mu\mu}$(eV)
C	−11·22	10·60	−11·16	11·13
N as in —N$=$	−14·51	13·31	−14·12	12·34
N as in —N—			−25·73	16·76
O as in $=$O	−17·25	14·67	−17·70	15·23
O as in —O—			−30·07	19·24

Little work has been done to evaluate the relative merits of the different sets of parameters we have described,[27–29] and some useful work could probably still be done in this field.

This completes our description of how the various quantities appearing in the Pople SCF equations (2.11) and (2.19), may be determined.

2.3   Application and development of the Pople π-SCF method

Most of the published work using the Pople SCF method has been concerned with the calculation of electronic spectra. As was stated earlier parameters chosen to fit excitation energies are generally unsuitable for the calculation of ground-state properties. Consequently we shall deal with the two types of data separately. We consider first the application of the Pople π-SCF method to the calculation of ground-state properties, notably heats of atomization.

Within the π-electron approximation it is assumed that the heat of atomization ΔH_a of a conjugated molecule is given by

$$\Delta H_a = E_\pi + E_\sigma \qquad (2.40)$$

where E_π, the total π-bond energy, is calculated by an SCF–MO treatment and E_σ is given by a sum of bond-additive contributions each of these being calculated from the appropriate Morse potential. In the case of the thermo-cycle method of Dewar and Schmeising[19] (2.31), the C—C σ-bond energy is automatically calculated during the evaluation of β. The X—H bond energies (X = C, N, O etc.) are usually taken as constant, because the bond lengths show little variation between the molecules. These energies are obtained from experimental data.

Chung and Dewar[30] were the first to develop an SCF–MO treatment for the ground-state properties of conjugated hydrocarbons and this was later extended to heteroconjugated systems by Dewar and Gleicher.[31] In this work the σ-bond energy was considered to be an independent parameter, but in subsequent work[20,21] the σ-bond energy was calculated directly from the thermocycle. This modification reduced the number of parameters in the theory and at the same time gave an improved agreement between theory and experiment. Table 2.3 shows heats of atomization calculated with this improved version for some conjugated molecules; the agreement with experiment is clearly very good.

TABLE 2.3 Comparison of calculated and observed heats of atomization (eV)[20,21]

Compound	Calc. ΔH_a	Obs. ΔH_a	Difference
Benzene	57·16	57·16	0·00
Naphthalene	90·61	90·61	0·00
Anthracene	123·89	123·93	−0·04
Azulene	89·46	89·19	0·27
Biphenyl	109·75	109·76	−0·01
Butadiene	42·05	42·05	0·00
Styrene	75·91	75·83	0·08
Triphenylene	157·94	157·76	0·18
Pyridine	51·87	51·79	0·08
Pyrazine	46·27	46·44	−0·17
Aniline	64·39	64·31	0·08
Imidazole	39·74	39·66	0·08
Furan	41·56	41·52	0·04
Benzaldehyde	68·51	68·27	0·24

Lo and Whitehead[32,33] and Loquet[34] have employed different parameter schemes within their SCF–MO treatments and have achieved results for hydrocarbons of comparable accuracy to that shown above.

Recently Dewar[35,36] has adopted the much simpler Mulliken expression (2.29) to evaluate β, because this was found to give similar results to those achieved by the thermocycle method. Expression (2.29) has the simplification that it contains only one parameter and it can therefore easily be extended to a wide variety of heteroconjugated systems, as a recent extension to sulphur[37] has shown. The thermocycle method, on the other hand, is not easily extended. A similar treatment in which β is evaluated from the Wolfsberg–Helmholtz expression (2.30) has also been applied to some sulphur conjugated molecules.[38]

The results shown in table 2.3 were obtained using the variable-β procedure of Dewar and Gleicher,[39] in which all bond lengths are re-calculated at each step in the iterative procedure from the bond-order expression (2.35), and the integrals $\beta_{\mu\nu}$ and $\gamma_{\mu\nu}$ between bonded atoms recalculated accordingly. Self-consistent bond lengths obtained by such a procedure are usually within $\pm 0 \cdot 02$ Å of the experimental bond lengths. Lo and Whitehead[32,33] use a more rigorous but more tedious procedure in which the total bond energy ($E_\pi + E_\sigma$) is minimized with respect to each C—C bond length.

Dewar, Hashmall and Venier[40] used the thermocycle method of para-meterization in SCF calculations of the heats of atomization of radicals, the parameters having been derived from the results on closed-shell mole-cules. However the radicals were then found to be too stable by approxi-mately $0 \cdot 5$ eV. It was suggested that this discrepancy arose from the fact that the SCF equations appropriate to open-shell molecules[41] take a greater account of electron correlation than the closed-shell equations. Thus to use closed-shell parameters in an open-shell treatment would lead to an overestimation of electron correlation. This led Dewar, Hashmall and Venier[40] to modify the closed-shell method for radicals by replacing the unpaired electron with two half-electrons of opposite spin. It can be shown that the total electronic energy of the radical is related to the energy calculated by this pseudo closed-shell method (E) by the expression

$$E_{\text{rad}} = E - \tfrac{1}{4} J_{00} \qquad (2.41)$$

where J_{00} is the repulsion energy for two electrons in the highest occupied molecular orbital. Heats of atomization calculated for radicals by this procedure are found to be accurate to within $\pm 0 \cdot 2$ eV.

We have therefore established that the Pople π-SCF method satisfactorily predicts ground-state properties. The situation is quite different however when we come to consider excited states, for here considerable mixing occurs between the different excited configurations so that the resulting stages cannot be represented by a single Slater determinant as can the ground state. However the wave functions can always be improved by writing them as linear combinations of Slater determinants.

$$\Psi = \sum_r c_r \Phi_r \tag{2.42}$$

where each Slater determinant Φ_r represents an electronic configuration of the molecule. The coefficients are determined by the variation method, that is by solving

$$|\mathcal{H}_{rs} - ES_{rs}| = 0 \tag{2.43}$$

to obtain the minimum total energy (see appendix 1). This procedure is known as the configuration interaction (CI) treatment.

One question that is posed by such a CI treatment is, how much CI needs to be considered to give reasonable results? Obviously we cannot hope to do a complete CI calculation on large molecules even given modern day computing facilities, because the number of such configurations obtained from even a minimum basis of atomic orbitals may be very large. An answer to this question has been provided by several authors[42–44] who conclude that at least those configurations which arise from the ground-state configuration by promoting one or two electrons, must be considered.

The Pople π-SCF method with configuration interaction has been extensively and successfully used in the interpretation of electronic spectra. Bibliography for this work has been given in Chapter 1 and we report only a few of the more recent calculations. An interesting addition to the already numerous SCF treatments of excited states has been provided recently by Bailey.[45] The important feature of her method is the use of the Linderberg expression (2.36) to calculate β. The treatment differs from conventional methods in that it contains no arbitrary parameters and yet gives transition energies which are in good agreement with experiment for hydrocarbons and heteroconjugated molecules. In table 2.4 we have summarized some of the results obtained by Bailey using the Linderberg expression for β and for comparison her results based on the Nishimoto–Forster expression (2.38). The experimental values quoted are from her paper. The method has since been used to calculate the absorption spectra of some purines and pyrimidines of biological interest.[46]

The above results were obtained with a fixed-β procedure. Nishimoto and Forster have used a variable-β procedure within an SCF treatment to calculate the absorption spectra of hydrocarbons[23] and heteroaromatic

TABLE 2.4 Comparison of observed and calculated excitation energies and oscillator strengths from the work of Bailey[45]

Compound	Singlet transition energies (eV)			Oscillator strengths		
	A	B	Exp.	A	B	Exp.
Benzene	4·85	4·89	4·89	0·0	0·0	0·01
	6·09	6·18	6·17	0·0	0·0	0·126
	6·92	7·01	6·98	0·55	1·19	1·04
Pyridine	4·89	4·98	4·96	0·02	0·06	0·03
	6·18	6·30	6·36	0·01	0·04	0·2
	7·05	7·24	7·04	0·49	1·13	1·36
	7·12	7·17		0·54	1·13	
Aniline	4·38	4·40	4·40	0·02	0·06	0·03
	5·29	5·38	5·39	0·20	0·40	0·14
	6·27	6·41	6·40	0·17	0·52	0·51
	6·59	6·48	6·88	0·36	0·90	0·57
	7·50	7·67	7·87	0·32	0·49	0·68
	7·60	7·62		0·02	0·01	
p-Benzoquinone	4·17	3·59	4·28	0·0	0·0	
	4·83	4·51	5·07	0·49	0·85	
	6·68	6·28		0·0	0·0	
	6·69	6·61		0·0	0·0	
	7·17	7·10		0·10	0·20	
Phenol	4·65	4·62	4·59	0·01	0·04	0·02
	5·72	5·75	5·78	0·10	0·21	0·18
	6·64	6·71	6·68	0·37	0·89	0·59
	6·70	6·78	6·90	0·47	1·12	0·37

A. Using Linderberg β (2·37).   B. Using Nishimoto–Forster β (2·38).

systems.[47] Their procedure is similar to that of Dewar and Gleicher,[39] though less rigorous because the two-centre electron repulsion integrals $\gamma_{\mu\nu}$ are not varied. However, their ground-state bond lengths agree closely with those obtained by Dewar and Gleicher because the range over which $\gamma_{\mu\nu}$ varies is not large. The advantage of the variable-β procedure is that it does not require a foreknowledge of the molecular geometry.

At the present time, to our knowledge, no variable-β procedure has been described in which the integrals between non-neighbour atoms are varied. One way of achieving this for the cyclic compounds would be to assume a relationship between bond angles and bond lengths in closed rings (e.g. one can assume that in a 6-membered ring the bond angles adjust to the bond lengths in such a way that the sum of squared deviations from the natural angle of 120° is a minimum).[48] It is doubtful however if this more rigorous procedure would lead to any significant improvement in results.

The variable-β procedure has been used in conjunction with the half-electron method[40] to calculate adiabatic ionization potentials; these are defined as the difference in energy between a molecule and its positive ion (a radical) when both are in their most stable nuclear configurations. Values for adiabatic ionization potentials are obtained experimentally from photoionization or photoelectron spectroscopy. Vertical ionization potentials, on the other hand, are defined as the difference in energy between a molecule in its ground state and the positive ion with the same molecular dimensions. Experimentally they are obtained from electron-impact measurements or from the Franck–Condon maxima in photoelectron spectra.

Two procedures are used to calculate vertical ionization potentials within the π-SCF method. The first procedure, and the most popular, makes use of Koopmans' theorem[24] and equates the vertical ionization potential to minus the corresponding orbital energy. The second procedure is to calculate directly the energy of a molecule using the variable β procedure, and then to calculate the cation with the same geometry as the parent molecule. These two procedures lead to almost identical vertical ionization potentials. In table 2.5 we list some vertical and adiabatic

TABLE 2.5   Comparison of calculated and observed ionization potentials

| | Adiabatic ionization potential (eV) | | Vertical ionization potential (eV) | | |
	Dewar[36,40]	Exp.	Nishimoto–Forster[23,47]	Dewar[36,40]	Exp.
Benzene	9·22	9·24	—	9·35	9·38
Naphthalene	8·30	8·12	8·23	8·45	8·26
Anthracene	7·72	7·38	7·65	7·83	7·55
Phenanthrene	8·10	7·91	8·17	8·28	8·03
Pyrene	7·71		7·67	7·97	7·72
Azulene	7·54	7·43	7·48	7·63	7·72
Pyridine	10·09	9·31	9·29		
Quinoline	8·56	8·62	8·62		
Isoquinoline	8·36	8·54	8·42		
Phenol	8·86	8·52	8·24		

ionization potentials obtained from the π-SCF-MO method of Dewar and coworkers together with the vertical ionization potentials obtained by the method of Nishimoto and Forster. Although both methods use quite different values for β, they predict rather similar ionization potentials. The reason for this is that both groups make an empirical correction to their orbital energies so that they can be equated to experimental ionization potentials. Thus Dewar and coworkers calculate the orbital energies relative to the ionization potential of the methyl radical (9·84 eV), whilst Nishimoto and Forster employ a correction so that the experimental ionization potential of benzene (9·25 eV) is reproduced. Electron affinities have been calculated in the same way[40,23,47] but because few experimental values are known accurately we have not listed them.

When the parameters optimized for ground-state properties are used to calculate excitation energies the results are usually much too high. This is true however only if the excited states are built up from SCF orbitals of the ground state and only a limited amount of configuration interaction is introduced. Dewar and Trinajstić[49] have shown that if new orbitals are calculated for excited configurations of the molecules and these are then used to form excited states, the excitation energies calculated from ground-state parameters are quite good. The SCF orbitals for these open-shell configurations were obtained using an extension of the half-electron technique used for radicals. Thus the SCF orbitals for the lowest triplet states are obtained by assuming that two orbitals each contain two half-electrons of opposite spin. Calculations of this type can only be used to obtain a limited number of sets of SCF orbitals, that is, those appropriate to a closed-shell system, or to two unpaired electrons or to four unpaired electrons, etc. It remains to be seen whether the SCF orbitals calculated for two unpaired electrons are a good basis for describing excited states other than the lowest triplet and singlet state.

Dewar and Trinajstić[49] use both a fixed ground-state geometry, to get vertical excitation energies, and use the variable-β procedure to obtain optimum excited state geometries and hence the 0–0 transition energy. The difference between the two may be as large as 0·5 eV. Table 2.6 shows

TABLE 2.6 Comparison of observed and calculated 0–0 transition energies (eV) to the L_a states[49]

	Singlet–triplet		Singlet–singlet	
	Calc.	Exp.	Calc.	Exp.
Naphthalene	2·06	2·62	4·22	4·30
Biphenyl	2·44	2·82	4·44	4·90
Anthracene	1·57	1·82	3·33	3·31
Phenanthrene	2·26	2·68	4·26	4·24
Triphenylene	2·76	2·89	4·28	4·32
Fluoranthene	1·94	2·29	3·56	3·46
Coronene	2·27	2·39	4·05	3·63
Styrene	2·04	2·68	4·76	5·06
Aniline	3·18	3·32	4·58	4·29
α-naphthylamine	2·05	2·49	4·17	3·87
Carbazole	2·67	3·04	4·57	4·27
Quinoline	2·10	2·71	4·02	4·26
Acridine	1·80	1·96	3·58	3·52
Phenol	3·57	3·53	4·97	4·59
α-naphthol	2·18	2·54	4·24	4·22
Benzophenone	3·54	2·99	5·22	4·90

some of their results, which compare favourably with those obtained by other workers.[50–52]

Several authors[53–56] have drawn attention to the fact that greater justification of the ZDO approximation can be obtained by using as a basis set, orthogonalized atomic orbitals. Fischer–Hjalmars[57] has carried out a detailed investigation of this problem and found that the conditions imposed by the ZDO approximation will be fulfilled to second-order in overlap in a basis of orbitals orthogonalized according to the so-called Löwdin recipe[58]

$$\{\psi\} = \mathbf{S}^{-\frac{1}{2}}\{\varphi\} \qquad (2.44)$$

Likewise McWeeny[59] has shown that when electron repulsion integrals are calculated using Löwdin orbitals then those integrals which are neglected in the ZDO approximation are, in fact, extremely small. This can clearly be seen from table 2.7 where we have listed some electron repulsion integrals which were calculated for the allyl radical using Löwdin

TABLE 2.7 Comparison of two-electron integrals between orthogonal and non-orthogonal orbitals for the allyl radical (energies in eV)

Integral	Non-orthogonal orbitals	Orthogonal orbitals
$(11 \mid 11)$	16·9341	17·2625
$(22 \mid 22)$	16·9300	17·6093
$(11 \mid 22)$	9·0295	8·8549
$(11 \mid 33)$	5·6703	5·5814
$(11 \mid 12)$	3·3130	−0·0931
$(12 \mid 12)$	0·9276	0·0973
$(12 \mid 13)$	0·1119	−0·0027

orbitals[60]; similar values were obtained for the electron repulsion integrals of benzene.[60] The one and two-centre coulomb integrals [$(11 \mid 11)$, $(11 \mid 22)$ etc.] are slightly changed by orthogonalization; thus any SCF–MO method which uses Löwdin orbitals as a basis set, would require different parameter values to those used in conventional SCF–MO methods. The low value of $\gamma_{\mu\mu}$ deduced from $I - A$ cannot be attributed to the effective orthogonalization of the atomic orbital basis implied in a ZDO scheme. The other significant point about the results of table 2.7 is that after orthogonalization γ_{11} is different from γ_{22}, and both are different from the γ_{11} found for benzene. Thus in an orthogonalized basis one would not strictly have integrals which are transferrable to all sp^2 hybridized carbon atoms.

Adams and Miller have described an SCF–MO treatment for conjugated systems containing carbon, nitrogen and oxygen in which Löwdin orbitals are used as a basis set.[61,62] Ionization potentials and singlet and triplet excitation energies calculated by this method were in satisfactory agreement with experiment. The method of Adams and Miller has since been extended by other workers to allow for variable β and a variable electronegativity procedure which will be described shortly.[63] A critical assessment of the method has been given[64] and a comparative study with other SCF–MO methods has been made.[65] The conclusions were that the original treatment of Adams and Miller did not lead to any real improvement over the more conventional type of SCF–MO method; singlet excitation energies are no better than those obtained by Bailey (see table 2.4). Recently Baird[66] has developed an SCF–MO treatment of hydrocarbons using the usual Slater-type orbital basis set, in which overlap integrals on neighbouring atoms are not neglected. This leads, it is claimed, to an improvement in calculated singlet excitation energies over comparable ZDO methods.

Some of the well-known properties of Hückel orbitals are retained by the ZDO–SCF orbitals of hydrocarbons. In particular, providing that only β between neighbouring atoms is considered, then the familiar pairing properties of the orbitals of alternant hydrocarbons are maintained,[67] together with the related property that neutral alternants have a uniform distribution of π electrons. Even for non-alternants one finds rather small net π charges on the atoms—smaller than those obtained from Hückel calculations.

ZDO calculations[68,69] in which resonance integrals between non-neighbouring atoms are also considered, lead to a non-uniform charge distribution in some alternant hydrocarbons notably those with 4-membered rings. It is doubtful whether the inclusion of these long range β's leads to significantly improved results, but the method has since been extended to nitrogen and oxygen heterocycles.[70]

Attempts have been made within the SCF–MO method to take a more explicit account of electron correlation. The most recent attempt was by Nishimoto[71] who allowed for the effect of 'vertical' electron correlation, i.e. correlation of electrons in a direction perpendicular to the molecular symmetry plane in π-electron molecules, by modifying the two-centre electron repulsion expression so as to distinguish between upper–upper and upper–lower electron interactions. This follows a similar procedure developed earlier by Dewar and Hojvat and called by them the split-p orbital (SPO) method.[72,73] A basic deficiency of the SPO method is that the orbitals are not orthogonal to the core[74–76] and thus do not satisfy a necessary condition for sigma–pi separability. Although such procedures have academic interest, it is debatable whether they are of any real practical value when electron correlation may be allowed for by choosing parameters to give the best possible agreement between theory and experiment.

The validity of the sigma–pi separability condition imposed in the Pople π-SCF method has been discussed[77] and recent *ab initio* calculations have shown it to be untenable particularly for molecules having polar bonds[78,79] Brown and Heffernan[80,81] were the first to take account of the electron distribution in the core through a variable electronegativity (VESCF) procedure, in which the effective nuclear charge Z_ρ is related to the nuclear charge N_ρ by the expression

$$Z_\rho = N_\rho - 1.35 - 0.35 \, (\sigma_\rho + P_{\rho\rho}) \tag{2.45}$$

σ_ρ is the number of σ electrons contributed to the core by the atom and $P_{\rho\rho}$ is the π-electron charge on the atom. Brown and Heffernan proposed that the valence-state ionization potential be related to the effective nuclear charge Z by a quadratic expression

$$I = A + BZ + CZ^2 \tag{2.46}$$

in which the parameters A, B and C are to be determined from spectroscopic data for an iso-electronic series. The one-centre electron repulsion integral $\gamma_{\rho\rho}$ was also related to Z_ρ following an expression due to Paoloni[82]

$$\gamma_{\rho\rho} = 3.294 Z_\rho \text{ (eV)}, \tag{2.47}$$

and the two-centre electron repulsion integral $\gamma_{\mu\nu}$ was calculated from a similar expression to that used by Pariser and Parr (2.25). Thus any changes in the σ core will be reflected in the integrals $\gamma_{\rho\rho}$, $\gamma_{\rho\sigma}$ and $U_{\rho\rho}$ since these are all functions of the effective nuclear charge Z_ρ. A VESCF procedure of excited states in which all non-neighbour β's are retained has recently been described.[83]

The VESCF method has been adopted in various forms for the calculation of excited-state[84–87,98] and ground-state properties[21,88]. Ross and Skancke[89] have allowed for the effect of core polarization by increasing $U_{\mu\mu}$ (i.e. to make it more positive) for a carbon atom when the number of bonded H atoms is decreased. This treatment which the authors claim leads to considerable improvement in electronic transitions and ionization potentials, has since been extended to heteroaromatic systems.[90]

π-electron theories suffer from the restriction that they can only be strictly applied to planar conjugated molecules. Moreover if there are steric effects associated with non-bonded interactions, as for example in butadiene, such a theory is unlikely to give a satisfactory energy unless these interactions are allowed for. However π-electron calculations have been carried out on non-planar conjugated molecules (e.g. biphenyl) to predict the most stable conformation. It was found[91] that a satisfactory potential function for rotation about a C—C bond could only be obtained if one considered the combined effect of conjugation and non-bonded interaction; one may see this intuitively since for most conjugated molecules the two effects are in opposition, that is conjugation favours the planar form and non-bonded interactions favour the non-planar form. The π-electronic energy of a bond A—B as a function of the twist angle θ, can be calculated within the π-SCF method by assuming the core resonance integral β to have the following dependence on θ

$$\beta_{AB} = \beta_{AB}{}^{\circ} \cos \theta \qquad (2.48)$$

where $\beta_{AB}{}^{\circ}$ is the resonance integral for the planar system; this is the only modification introduced into the method. Non-bonded interactions are calculated from empirical potential functions, which relate the interaction energy to the distance r between non-bonded atoms. The behaviour of some of these functions has been studied by Fischer–Hjalmars[91] who adopted, like others, the following expressions of Bartell[92]

$$V_{HH} = 0{\cdot}659 \times 10^4 \exp(-4{\cdot}082r) - 0{\cdot}492 \times 10^2 r^{-6} \qquad (2.49)$$

$$V_{CH} = 1{\cdot}2496 \times 10^2 r^{-6} [358 \exp(-2{\cdot}041r) - 1] \qquad (2.50)$$

$$V_{CC} = 2{\cdot}994 \times 10^5 r^{-12} - 3{\cdot}253 \times 10^2 r^{-6}. \qquad (2.51)$$

These expressions give energies in kcal/mole if r is in Å.

Two molecules which have been extensively studied and which provide a rather crucial test of the theory are biphenyl and butadiene. Biphenyl is an interesting molecule, being planar in the solid state[93] and non-planar in the gas phase ($\theta = 42°$).[94] In the solid state the steric repulsion is partly relieved by distortion of the benzene rings.[95] Butadiene exists in the planar trans-conformation.[96]

It is appropriate at this point to mention that because solvation and crystal-field forces are neglected in the SCF–MO method, then clearly the SCF results obtained from such a method refer to the gas-phase. Although this distinction is irrelevant for most of the cases treated with the π-SCF method, there are a few crucial cases in which the point seems to have been overlooked.

The π-SCF–MO method in conjunction with the Bartell expressions successfully predicts biphenyl to be non-planar ($\theta = 40°$) and butadiene to exist in a planar trans-conformation; calculated barriers to internal rotation are in less satisfactory agreement with experiment.[33,36,91] In a similar manner π-SCF methods have been used in combination with appropriate steric potentials to calculate the relative rates of reaction (proton abstraction and solvolysis) of some sterically hindered arylmethyl compounds.[97] However, even given this partial success, the π-SCF–MO method is still rather limited in its application. Furthermore if it is accepted that π-electron energy levels depend on the σ-electron distribution, as allowed for in the VESCF procedure, then one must assume that a reciprocal relation applies, σ-energy levels depend on the π-electron distribution. At this level of complication the problem would become simpler by using one of the all-electron theories which will be described in the next chapter.

References

1. R. Pariser and R. G. Parr, *J. Chem. Phys.*, **21**, 466 (1953).
2. R. Pariser and R. G. Parr, *J. Chem. Phys.*, **21**, 767 (1953).
3. J. A. Pople, *Trans. Faraday Soc.*, **49**, 1375 (1953).
4. J. C. Slater, *Quantum Theory of Atomic Structure*, McGraw-Hill, New York, 1960.
5. R. Pariser, *J. Chem. Phys.*, **21**, 568 (1953).
6. R. G. Parr, D. P. Craig and I. G. Ross, *J. Chem. Phys.*, **18**, 1561 (1950).
7. W. Moffitt, *Proc. Roy. Soc. (London)*, **A210**, 224 (1951).
8. H. Sponer and P. O. Löwdin, *J. Physique Radium*, **15**, 607 (1954).
9. M. K. Orloff and O. Sinanoglu, *J. Chem. Phys.*, **43**, 49 (1965).
10. T. Anno, *J. Chem. Phys.*, **47**, 5335 (1967).
11. R. B. Hermann, *J. Chem. Phys.*, **42**, 1027 (1965).
12. R. G. Parr, *J. Chem. Phys.*, **20**, 1499 (1952).
13. N. Mataga and K. Nishimoto, *Z. Physik Chem. (Frankfurt)*, **12**, 335; **13**, 140 (1957).
14. K. Ohno, *Theoret. Chim. Acta*, **2**, 219 (1964).
15. G. Klopman, *J. Amer. Chem. Soc.*, **86**, 4550 (1964).
16. J. Koutecký, *J. Chem. Phys.*, **47**, 1501 (1967).
17. R. S. Mulliken, *J. chim. phys.*, **46**, 497, 675 (1949).
18. M. Wolfsberg and L. Helmholtz, *J. Chem. Phys.*, **20**, 837 (1952).
19. M. J. S. Dewar and H. N. Schmeising, *Tetrahedron*, **5**, 166 (1959), **11**, 96 (1960).
20. M. J. S. Dewar and C. de Llano, *J. Amer. Chem. Soc.*, **91**, 789 (1969).
21. M. J. S. Dewar and T. Morita, *J. Amer. Chem. Soc.*, **91**, 796 (1969).
22. J. Linderberg, *Chem. Phys. Letters*, **1**, 39 (1967).
23. K. Nishimoto and L. S. Forster, *Theoret. Chim. Acta* **3**, 407 (1965).
24. T. Koopmans, *Physica*, **1**, 104 (1934).
25. G. Pilcher and H. A. Skinner, *J. Inorg. Nucl. Chem.*, **24**, 937 (1962).
26. J. Hinze and H. H. Jaffé, *J. Amer. Chem. Soc.*, **84**, 540 (1962).
27. R. L. Flurry and J. J. Bell, *Theoret. Chim. Acta*, **10**, 1 (1968).

28. F. B. Billingsley and J. E. Bloor, *Theoret. Chim. Acta*, **11**, 325 (1968).
29. P. Chiorboli, A. Rastelli and F. Momicchioli, *Theoret. Chim. Acta*, **5**, 1 (1966).
30. A. L. H. Chung and M. J. S. Dewar, *J. Chem. Phys.*, **42**, 756 (1965).
31. M. J. S. Dewar and G. J. Gleicher, *J. Chem. Phys.*, **44**, 759 (1966).
32. D. H. Lo and M. A. Whitehead, *Can. J. Chem.*, **46**, 2027 (1968).
33. D. H. Lo and M. A. Whitehead, *Can. J. Chem.*, **46**, 2041 (1968).
34. A. J. Lorquet, *Theoret. Chim. Acta*, **5**, 192 (1966).
35. M. J. S. Dewar and A. J. Harget, *Proc. Roy. Soc.*, **A315**, 443 (1970).
36. M. J. S. Dewar and A. J. Harget, *Proc. Roy. Soc.*, **A315**, 457 (1970).
37. M. J. S. Dewar and N. Trinajstic, *J. Amer. Chem. Soc.*, **92**, 1453 (1970).
38. D. T. Clark, *Tetrahedron*, **24**, 2567 (1968).
39. M. J. S. Dewar and G. J. Gleicher, *J. Amer. Chem. Soc.*, **87**, 685 (1965).
40. M. J. S. Dewar, J. A. Hashmall and C. G. Verrier, *J. Amer. Chem. Soc.*, **90**, 1953 (1968).
41. A. Brickstock and J. A. Pople, *Trans. Faraday Soc.*, **50**, 901 (1954).
42. N. L. Allinger and T. W. Stuart, *J. Chem. Phys.*, **47**, 4611 (1967).
43. J. Koutecký, K. Hlavatý and P. Hochmann, *Theoret. Chim. Acta*, **3**, 341 (1965).
44. E. M. Evleth, *J. Chem. Phys.*, **46**, 4151 (1967).
45. M. L. Bailey, *Theoret. Chim. Acta*, **13**, 56 (1969).
46. M. L. Bailey, *Theoret. Chim. Acta*, **16**, 309 (1970).
47. K. Nishimoto and L. S. Forster, *Theoret. Chim. Acta*, **4**, 155 (1966).
48. R. D. Brown and B. A. W. Coller, *Theoret. Chim. Acta*, **7**, 259 (1967).
49. M. J. S. Dewar and N. Trinajstić, *J. Chem. Soc.*, A, 1220 (1971).
50. M. K. Orloff, *J. Chem. Phys.*, **47**, 235 (1967).
51. K. Nishimoto and L. S. Forster, *J. Chem. Phys.*, **47**, 5451 (1967).
52. J. Pancír and R. Zahradnik, *Theoret. Chim. Acta*, **14**, 426 (1969).
53. P. O. Löwdin, *Advances in Physics*, **5**, 111 (1956).
54. R. McWeeny, *Proc. Roy. Soc.*, **A227**, 288 (1955), **A237**, 355 (1956).
55. K. Ruedenberg, *J. Chem. Phys.*, **34**, 1861 (1961).
56. P. G. Lykos, *J. Chem. Phys.*, **35**, 1249 (1961).
57. I. Fischer-Hjalmars, *J. Chem. Phys.*, **42**, 1962 (1965); *Advan. Quantum Chem.*, **2**, 25 (1965).
58. P. O. Löwdin, *J. Chem. Phys.*, **18**, 365 (1950).
59. R. McWeeny, *M.I.T. Solid State and Molecular Group, Quarterly Progress Report*, **11**, 25 (1954).
60. D. P. Chong, *Mol. Phys.*, **10**, 67 (1965).
61. O. W. Adams and R. L. Miller, *J. Amer. Chem. Soc.*, **88**, 404 (1966).
62. O. W. Adams and R. L. Miller, *Theoret. Chim. Acta*, **12**, 151 (1968).
63. K. D. Warren and J. R. Yandle, *Theoret. Chim. Acta*, **12**, 279 (1968).
64. K. D. Warren and J. R. Yandle, *Theoret. Chim. Acta*, **12**, 267 (1968).
65. J. E. Bloor, B. R. Gilson and N. Brearley, *Theoret. Chim. Acta*, **8**, 35 (1967).
66. N. C. Baird, *Mol. Phys.*, **18**, 39 (1970).
67. C. A. Coulson and G. A. Rushbrooke, *Proc. Camb. Phil. Soc. Math. Phys. Sci.*, **36**, 193 (1940).
68. R. D. Brown, F. R. Burden and G. M. Mohay, *Aust. J. Chem.*, **21**, 1695 (1968).
69. R. L. Flurry and J. J. Bell, *J. Amer. Chem. Soc.*, **89**, 525 (1967).

70. R. L. Flurry, E. W. Stout and J. J. Bell, *Theoret. Chim. Acta*, **8**, 203 (1967).
71. K. Nishimoto, *Theoret. Chim. Acta*, **5**, 74 (1966); **7**, 207 (1967).
72. M. J. S. Dewar and N. L. Hojvat, *J. Chem. Phys.*, **34**, 1232 (1961).
73. M. J. S. Dewar and N. L. Hojvat, *Proc. Roy. Soc.*, **A264**, 431 (1961).
74. L. C. Snyder and R. G. Parr, *J. Chem. Phys.*, **34**, 1661 (1961).
75. J. S. Griffith, *J. Chem. Phys.*, **36**, 1689 (1962).
76. M. J. S. Dewar, *J. Chem. Phys.*, **36**, 1689 (1962).
77. P. G. Lykos and R. G. Parr, *J. Chem. Phys.*, **24**, 1166 (1956); **25**, 1301 (1956).
78. E. Clementi, *Chem. Rev.*, **68**, 341 (1968).
79. H. Preuss and G. Diercksen, *Internat. J. Quantum Chem.*, **1**, 349, 357, 361 (1967).
80. R. D. Brown and M. L. Heffernan, *Trans. Faraday Soc.*, **54**, 757 (1958).
81. R. D. Brown and M. L. Heffernan, *Aust. J. Chem.*, **12**, 319 (1959).
82. L. Paoloni, *Nuovo Cimento*, **4**, 410 (1956).
83. R. D. Brown, F. R. Burden and G. R. Williams, *Aust. J. Chem.*, **21**, 1939 (1968).
84. O. Matsuoka and Y. I'Haya, *Mol. Phys.*, **8**, 455 (1964).
85. D. T. Clark and J. W. Emsley, *Mol. Phys.*, **12**, 365 (1967).
86. N. L. Allinger, J. C. Tai and T. W. Stuart, *Theoret. Chim. Acta*, **8**, 101 (1967).
87. K. Nishimoto, *Theoret. Chim. Acta*, **10**, 65 (1968).
88. J. W. Emsley, *J. Chem. Soc.*, **A**, 2523 (1968).
89. B. Roos and P. N. Skancke, *Acta Chem. Scand.*, **21**, 233 (1967).
90. A. Skancke and P. N. Skancke, *Acta Chem. Scand.*, **24**, 23 (1970).
91. I. Fischer-Hjalmars, *Tetrahedron*, **19**, 1805 (1963).
92. L. S. Bartell, *J. Chem. Phys.*, **32**, 827 (1960).
93. J. Trotter, *Acta Cryst.*, **14**, 1135 (1961).
94. A. Almenningen and O. Bastiansen, *Kgl. Norske Vid. Selsk. Skr.*, **1958**, 4.
95. K. Miller and J. N. Murrell, *Trans. Faraday Soc.*, **63**, 806 (1967).
96. A. Almenningen, O. Bastiansen and M. Traetteberg, *Acta Chem. Scand.*, **12**, 1221 (1958).
97. G. J. Gleicher, *J. Amer. Chem. Soc.*, **90**, 3397 (1968).
98. J. C. Tai and N. L. Allinger, *Theoret. Chim. Acta*, **15**, 133 (1969).

Chapter 3

The theory and development of ZDO-SCF *all-valence-electron* methods

3.1 Problems in applying the zero-differential-overlap (ZDO) approximation to σ electrons

There was a considerable time interval between the birth of the π-electron theories described in the last chapter and the development of similar theories for all-valence electrons. There are probably several reasons for this which no doubt influenced workers in the field to different extents: the following seem to us to be the most likely.

1. For conjugated hydrocarbons the π-electron theories worked quite well and there did not appear to be any need to take account of sigma electrons.

2. The saturated hydrocarbons have rather dull electronic spectra and their heats of formation can generally be fitted by any sensible theory with two or three adjustable parameters (there were many empirical MO theories of saturated hydrocarbons which gave good bond energies). It was therefore difficult to find critical data on which to deduce the large number of parameters encountered in a ZDO–SCF theory of all-valence electrons.

3. The most critical compounds for testing a theory are those with several different types of atoms, but these had been difficult to parameterize even in π-electron theory.

4. The number of valence electrons even in small polyatomic molecules is sufficiently large to require the use of a digital computer in any SCF calculation. The π-electron calculations however could be done, with effort, on hand calculators.

5. The ZDO model has greater justification in a π-electron theory than in a σ-electron theory because the overlap integral between neighbouring π atomic orbitals is much smaller ($\sim 0 \cdot 25$) than the overlap integral between neighbouring σ orbitals (usually $> 0 \cdot 5$).

We stated earlier that one can justify the ZDO π-model by specifying that the orbitals used for the LCAO expansion (1.1) are orthogonalized atomic orbitals, and this justification is obviously even more necessary if the ZDO approximations are to be used for σ electrons. However, the transformation (2.44) does not produce a set of equivalent orthogonalized orbitals unless there are symmetry reasons for this (e.g. orthogonalized π orbitals of benzene). When overlap is small the nonequivalence is not large enough for it to be necessary to adopt different values for integrals like $U_{\mu\mu}$ and $\gamma_{\mu\mu}$ for different carbon atoms in a molecule like naphthalene. Cook and McWeeny[1] have shown that a ZDO theory for σ orbitals can be justified on the basis of orthogonalized orbitals, but that these orbitals are rather far removed from the original non-orthogonal orbitals.

In 1965, Pople, Santry and Segal[2] published an important paper in which they showed that if one introduced ZDO approximations into an all-valence-electron theory then certain conditions had to be satisfied for the resulting energies and electron densities to be invariant to the choice of axes and hybridization of the basis set. In this paper they described two limiting approximations, the Complete Neglect of Differential Overlap (CNDO) approximation and the more rigorous, Neglect of Diatomic Differential Overlap (NDDO) approximation. We shall describe first the CNDO method.

3.2 Complete neglect of differential overlap—CNDO

If the full SCF equations are solved without any approximations then the calculated energies and electron distribution will be the same whatever the choice of coordinate axes. In other words, it does not matter how we choose to direct our angle-dependent orbitals (p, d etc). The results must also be the same whether we choose to take a linear combination of atomic orbitals, or a linear combination of hybridized orbitals which are themselves linear combinations of atomic orbitals. We say that the results of the SCF calculation are invariant to an orthogonal transformation of the atomic orbital basis. If one makes approximations to the SCF equations then the conditions of rotational and hybridizational invariance may not necessarily be satisfied. This problem did not arise in π-electron theory as the molecular symmetry leads to a natural choice of axis for the π orbitals, namely perpendicular to the molecular plane. However, for a molecule with no symmetry one would clearly not want the results to depend on the choice of axis when none is obvious.

The restrictions that the above invariance conditions impose on the approximations that can be made to the SCF equations, can be seen more clearly if we consider a typical two-electron integral of the type that is retained in a ZDO theory

$$(p_A^2 \mid s_B^2). \tag{3.1}$$

Suppose now that p_A is directed along a vector in the xy plane, so that it can be compounded into px_A and py_A components as follows

$$p_A = \cos\theta\, px_A + \sin\theta\, py_A. \tag{3.2}$$

Introducing this into (3.1), the integral becomes

$$((\cos\theta\, px_A + \sin\theta\, py_A)^2 \mid s_B^2) = \cos^2\theta(px_A^2 \mid s_B^2) + \sin^2\theta(py_A^2 \mid s_B^2)$$
$$+ 2\sin\theta\cos\theta(px_A py_A \mid s_B^2) \tag{3.3}$$

The final integral in (3.3) would be neglected in a ZDO model because px and py are orthogonal. In this approximation we therefore require

$$(p_A^2 \mid s_B^2) = \cos^2\theta(px_A^2 \mid s_B^2) + \sin^2\theta(py_A^2 \mid s_B^2) \tag{3.4}$$

and for this equality to hold for all θ, that is for any choice of coordinate axes, we must have

$$(px_A^2 \mid s_B^2) = (py_A^2 \mid s_B^2) = (p_A^2 \mid s_B^2) \tag{3.5}$$

By extending this argument one can see that the integral $(p_A^2 \mid s_B^2)$ must be independent of the orientation of the p_A orbital. Furthermore, since the combination $(px^2 + py^2 + pz^2)$ has spherical symmetry then it must be equal to $(s_A'^2 \mid s_B^2)$ where s_A' is the spherical orbital with the same radial function as the p_A orbital. If in addition one requires that the ZDO model shall be invariant to hybridization of the basis set, then one can see from arguments similar to those given above that $(s_A'^2 \mid s_B^2)$ must equal $(s_A^2 \mid s_B^2)$.

The SCF approximation in which all integrals $(\mu\nu \mid \rho\sigma)$ are neglected unless $\mu = \nu$ and $\rho = \sigma$, is called CNDO (complete neglect of differential overlap). For invariance to orthogonal transformations of the basis we then require that integrals like $(\mu^2 \mid \rho^2)$ are the same for all valence orbitals μ on atom M and ρ on atom A, and we put this integral equal to $(s_M^2 \mid s_A^2)$ for which we can use the symbol γ_{MA}.

Invariance conditions must also be applied to the one-electron integrals. For example, if we make the transformation (3.2) on a typical overlap integral, then we obtain

$$\int p_A s_B \, dv = \cos \theta \int p x_A s_B \, dv + \sin \theta \int p y_A s_B \, dv. \quad (3.6)$$

In the CNDO method the resonance integrals are taken to be proportional to the overlap integrals as in expression (2.29). In order that the resonance integrals are invariant to rotation we require

$$\beta_{p_A s_B} = \cos \theta \, \beta_{p x_A s_B} + \sin \theta \, \beta_{p y_A s_B}. \quad (3.7)$$

Introducing (2.29) on both sides of this equation we obtain

$$\beta^0_{p_A s_B} S_{p_A s_B} = \cos \theta \, \beta^0_{p x_A s_B} S_{p x_A s_B} + \sin \theta \, \beta^0_{p y_A s_B} S_{p y_A s_B}. \quad (3.8)$$

If equation (3.6) is true then (3.8) will only be true for all θ if β^0 has the same value in all cases, that is, it is independent of the orientation of the p orbital. If we combine this result with an invariance to hybridization we see that β^0 must be a constant depending only on the nature of the two atoms involved.

The expressions for the F-matrix elements in the CNDO approximation have a similar form to those for the π-electron model. From expression (2.11) we have

$$F_{\mu\nu} = \beta^0{}_{MN} S_{\mu\nu} - \tfrac{1}{2} P_{\mu\nu} \gamma_{MN}. \quad (3.9)$$

When μ and ν are on different atoms (M $\neq$ N), expression (3.9) has the same significance as in the π model. However, the two orbitals can also be on the same atom (M = N), in which case $S_{\mu\nu} = 0$ by virtue of their orthogonality, and the integral γ_{MN} is replaced by γ_{MM}; this case was not encountered in π-electron theory. For different orbitals on the same atom $P_{\mu\nu}$ will not necessarily be zero unless there are symmetry reasons for it to be so, and therefore the second term in (3.9) will contribute in this case to $F_{\mu\nu}$.

For the diagonal elements one can start from expression (2.12) but note that orbital φ_ρ may be on the same atom as φ_μ or on a different atom (A $\neq$ M)

$$F_{\mu\mu}{}^M = H_{\mu\mu}{}^c + \tfrac{1}{2} P_{\mu\mu} \gamma_{\mu\mu} + \sum_{\rho(M)} P_{\rho\rho} \gamma_{\mu\rho} + \sum_{\rho(A)}{}' P_{\rho\rho} \gamma_{\mu\rho}. \quad (3.10)$$

As the $\gamma_{\mu\rho}$ depend only on the nature of the two atoms, we can combine together the sums over ρ on the same atom, and introduce a net atom charge by

$$\sum_{\rho(A)} P_{\rho\rho} = P_{AA}. \quad (3.11)$$

The core matrix elements can also be split up as in 2.14, to give the final expression.

$$F_{\mu\mu}{}^{M} = U_{\mu\mu} - \tfrac{1}{2}P_{\mu\mu}\gamma_{MM} + P_{MM}\gamma_{MM} + \sideset{}{'}\sum_{A}(P_{AA}\gamma_{MA} - V_{A,\mu\mu}). \quad (3.12)$$

In the original CNDO method, called by Pople and Segal[3] CNDO/1, the integrals $V_{A,\mu\mu}$ and γ_{MA} were evaluated separately using Slater orbitals. In a later paper[4] they introduced two modifications to their theory. The first was in the method of evaluating the integral $U_{\mu\mu}$, and we shall have more to say about this shortly. The second modification was to use expression (3.12) with zero penetration ($f = 1$ in 2.18) so that now the expression for $F_{\mu\mu}$ takes on a similar form to expression (2.19)

$$F_{\mu\mu}{}^{M} = U_{\mu\mu} - \tfrac{1}{2}P_{\mu\mu}\gamma_{MM} + P_{MM}\gamma_{MM} + \sideset{}{'}\sum_{A}(P_{AA} - Z_{A})\gamma_{MA}. \quad (3.13)$$

This modified version was called by Pople and Segal[4] CNDO/2 and has proved to be the more popular version.

3.3   Choice of parameters for CNDO/2

The critical stage in the parameterization of any semi-empirical SCF method is in the evaluation of the integrals $U_{\mu\mu}$, and γ_{AA}; this is because these parameters determine the energy levels of the separate atoms, and the molecular energies will not be of the right order of magnitude unless the atomic energies are approximately correct. This is particularly true of the CNDO method which represents the simplest approximation proposed by Pople, Santry and Segal.[2] Various methods have been described to evaluate these integrals, most of which make use of the techniques introduced in the Pople π-SCF method which we have described in chapter 2.

The integral $U_{\mu\mu}$ represents the energy of a single electron occupying an orbital φ_{μ} in the field of the core; it is usually estimated from spectroscopic data. If the CNDO method is used to calculate the energy levels of an atom one finds that all the terms arising from a given atomic configuration have the same energy. This is because all one-centre exchange integrals are taken to be zero, and all the coulomb integrals have the same value. Thus the CNDO atomic parameters should only be obtained from the *average* energy of a configuration. The ground state of carbon, for example, has the electron configuration $2s^{2}2p^{2}$. This gives rise to three spectral terms ^{3}P, ^{1}D, and ^{1}S with energies 0, 1·263 and 2·683 eV respectively. The fact that these energies are different shows up an important deficiency of the CNDO method, which is corrected in the INDO method to be described later.

To find the average energy of a configuration one weights each term by its total degeneracy (spin multiplied by orbital degeneracy), so that for the ground configuration of carbon we have

$$\bar{E}(C, 2s^2 2p^2) = \tfrac{1}{15}(9E(^3P) + 5E(^1D) + E(^1S)) \tag{3.14}$$

In the CNDO method this is equal to

$$2U_{ss} + 2U_{pp} + 6\gamma_{CC} \tag{3.15}$$

there being six sets of electron pair repulsion for four electrons. The following more general expression is obtained, within the CNDO approximation, for the energy of a given configuration relative to the core state (i.e. the energy of the core state is set to zero).

$$\bar{E}(M, 2s^m 2p^n) = mU_{ss} + nU_{pp} + \tfrac{1}{2}(m + n)(m + n - 1)\gamma_{MM} \tag{3.16}$$

Suppose that we now remove an electron from a carbon $2p$ orbital. The resulting positive ion will have an energy

$$\bar{E}(C^+, 2s^2 2p^2) = 2U_{ss} + U_{pp} + 3\gamma_{CC}. \tag{3.17}$$

The difference between expressions (3.15) and (3.17) is equal to the ionization potential of a p electron

$$I_p = \bar{E}(C^+, 2s^2 2p) - \bar{E}(C, 2s^2 2p^2) = -U_{pp} - 3\gamma_{CC}. \tag{3.18}$$

In the CNDO/1 method, Pople and Segal evaluated the integral $U_{\mu\mu}$ from expression (3.18), with γ_{CC} being calculated theoretically using Slater s orbitals. This method was later modified in the CNDO/2 procedure on the grounds that $U_{\mu\mu}$ should be a measure not only of the difficulty of removing an electron from φ_μ but also of the ease of gaining one. Thus $U_{\mu\mu}$ was determined from a mean of the ionization potential and electron affinity according to the expression

$$U_{\mu\mu} = -\tfrac{1}{2}(I_\mu + A_\mu) - (Z_M - \tfrac{1}{2})\gamma_{MM} \tag{3.19}$$

where Z_M is the core charge of atom M. However, in this approach one encounters the difficulty that electron affinities are not reliably known for most atoms. Pople and Segal use a rather complicated method for their estimation based on the extrapolation of an iso-electronic series such as C^{2+}, B^+ and Be to Li^-. The resulting values of $\tfrac{1}{2}(I + A)$ for the first row atoms are shown in table 3.1

TABLE 3.1   Pople–Segal electronegativities used in the evaluation of $U_{\mu\mu}$ (eV)

	H	Li	Be	B	C	N	O	F
$\tfrac{1}{2}(I_s + A_s)$	7·176	3·106	5·946	9·594	14·051	19·316	25·390	32·272
$\tfrac{1}{2}(I_p + A_p)$	—	1·258	2·563	4·001	5·572	7·275	9·111	11·080

With the exception of hydrogen the ionization potentials quoted in table 3.1 were the same as used in the CNDO/1 method. The value $U_{\mu\mu}$ for hydrogen was obtained from the *calculated* electron affinity (0·747 eV) and the *experimental* ionization potential of 13·605 eV. In the CNDO/1 method the ionization potential for hydrogen was calculated for a 1s orbital with exponent 1·2, because such a contracted hydrogen orbital is frequently used in valence calculations.

When expression (3.19) is inserted into (3.13), $F_{\mu\mu}$ has the form

$$F_{\mu\mu} = -\tfrac{1}{2}(I_\mu + A_\mu) + [(P_{MM} - Z_M) - \tfrac{1}{2}(P_{\mu\mu} - 1)]\gamma_{MM}$$
$$+ \sum_{A \neq M} (P_{AA} - Z_A)\gamma_{MA} \quad (3.20)$$

For the case where all atoms are neutral, $P_{AA} = Z_A$, and orbital φ_μ contains one electron, (3.20) simplifies to

$$F_{\mu\mu} = -\tfrac{1}{2}(I_\mu + A_\mu) \quad (3.21)$$

which shows the relationship of $F_{\mu\mu}$ to the Mulliken electronegativity, $\tfrac{1}{2}(I + A)$, for the element.

Sichel and Whitehead[5] have adopted a different procedure to Pople and Segal in which the atomic parameters U_{ss}, U_{pp} and $\gamma_{\mu\mu}$ are evaluated from the valence-state data of Hinze and Jaffé.[6] The valence-state energy of an atom relative to the energy of the core is[6a]

$$E = \sum_\mu n_\mu U_{\mu\mu} + \tfrac{1}{2}\sum_\mu \sum_{\nu \neq \mu} n_\mu n_\nu \gamma_{\mu\nu} + \tfrac{1}{2}\sum_\mu n_\mu(n_\mu - 1)\gamma_{\mu\mu} \quad (3.22)$$

where n_μ is the orbital occupancy of orbital μ. Whenever possible Sichel and Whitehead chose data from valence states of low promotional energy and unit charge. The Pariser approximation (2.20) was adopted for the evaluation of the integrals $\gamma_{\mu\mu}$. For every atom with atomic number greater than five there will be four electron repulsion terms γ_{ss}, γ_{sp}, γ_{pp} and $\gamma_{pp'}$, the last term referring to two different valence p orbitals on the same atom. Thus for any atom there will be seven unknowns, the four electron-repulsion terms, the integrals U_{ss} and U_{pp} and finally the energy of the core. One therefore needs to know the energies of seven valence states (including states for the neutral atom and of its ions) to obtain values for these parameters. As stated earlier, one of the requirements of the CNDO approximation is that the one-centre electron-repulsion integrals γ_{AA} should have an average value characteristic of the atom A. Sichel and Whitehead in their determination therefore take γ_{AA} as an average of the four possible repulsion terms. Table 3.2 shows values obtained by them for U_{ss}, U_{pp} and γ_{AA} for the first-row atoms. They also give parameters

TABLE 3.2 The parameters of Sichel and Whitehead (eV)[5]

	H	Li	Be	B	C	N	O	F
U_{ss}	−13·595	−4·999	−15·543	−30·371	−50·686	−70·093	−101·306	−129·544
U_{pp}	—	−3·673	−12·280	−24·702	−41·530	−57·848	−84·284	−108·933
γ_{AA}	12·848	3·458	5·953	8·048	10·333	11·308	13·907	15·233

for the main group elements up to the fourth row of the periodic table, but not for the transition metals.

The values of U_{ss} and U_{pp} obtained by Sichel and Whitehead give better agreement with ionization potentials and electron affinities of the atoms than do the corresponding values of Pople and Segal. This result is not surprising because Sichel and Whitehead effectively introduce a correction for electron correlation by using the Pariser approximation. The values of γ_{AA} as calculated by Pople and Segal from Slater orbitals are much larger than those deduced from the Pariser approximation. Moreover, it was found by Sichel and Whitehead that γ_{ss} was the largest of the four electron-repulsion integrals, thus a calculated γ_{ss} is quite unrepresentative of the average electron-repulsion integral. For these reasons we think that the parameters of Sichel and Whitehead are to be preferred.

The constants β^0_{MN} occurring in the resonance integrals (3.9) were taken by Pople and Segal[3] to be the mean of constants β^0_M and β^0_N which were characteristic of the atoms M and N

$$\beta^0_{MN} = \tfrac{1}{2}(\beta^0_M + \beta^0_N). \qquad (3.23)$$

This is a simplification which slightly reduces the number of empirical parameters in the theory. The constants β^0_M were then chosen to give the best possible overall agreement between CNDO and non-empirical SCF calculations for some diatomic molecules: the resulting values are shown in table 3.3

Note that the values for β^0_M increase with increasing electronegativity, a fact which was built into the Wolfsberg–Helmolz approximation (2.30). This approximation has in fact been used in CNDO calculations to evaluate β but it does not satisfy the requirement of invariance to hybridization.

TABLE 3.3 Constants used to parameterize β in the CNDO/2 method

	H	Li	Be	B	C	N	O	F	
$-\beta^0_M$	9	9	13	17	21	25	31	39	(eV)

The two-centre electron repulsion terms γ_{AB} can be evaluated theoretically, as was done by Pople and Segal in their CNDO treatment. However this procedure will lead, for the reasons stated in chapter 2, to values for the integral γ_{AB} which are substantially greater than the values obtained semi-empirically. It would be possible to calculate γ_{AB} theoretically using orbital exponents which reproduce the semi-empirical values. A procedure of this kind was adopted by Miller and coworkers[7] in π-electron calculations, but the approach has gained little acceptance because it requires considerably more computer time than the semi-empirical evaluation of γ_{AB}. Thus the most widely adopted procedure has been to evaluate γ_{AB} from one of the empirical expressions (2.23 to 2.28) so successfully used in π-SCF calculations; of these the Mataga–Nishimoto approximation (2.26) has been the most frequently used for CNDO calculations other than those carried out by Pople and coworkers.

Finally, the repulsion energy of the cores was taken as the repulsion of nuclei with effective charges Z_A. This underestimates the repulsion at small internuclear distances and as a result equilibrium bond lengths calculated by the CNDO procedure are too short.

Of the various methods described so far to evaluate the integrals, we would recommend the procedure of Sichel and Whitehead for $U_{\mu\mu}$ and γ_{AA}, and either the Mataga–Nishimoto (2.26) or Ohno–Klopman (2.28) expressions for γ_{AB}. The final choice must be determined by the agreement between theory and experiment. Some of the properties studied with the CNDO method will now be described.

3.4   Applications of the CNDO method

In their original work on the CNDO method,[3,4] Pople and Segal aimed to reproduce the results obtained from *ab initio* calculations, and chose the parameters β^0_M accordingly. In general the results were encouraging. Orbital energies and total binding energies, were more negative than those obtained from the full SCF calculations, although the differences in orbital energies were well reproduced. Both the CNDO/1 and CNDO/2 methods gave equilibrium bond lengths which were considerably shorter than the observed values; the CNDO/2 method gave the better bond lengths and Segal[8] attributed this to the neglect of the penetration integrals.

Calculated equilibrium bond angles were in good agreement with experiment (e.g. H_2O calculated 107.1°, observed 104·5°; NH_3 calculated 106·7°, observed 106·6°) although bending-force constants were in general about 50% too high.[4] When one considers the small amount of energy required to change the bond angle in water from 104° to 90° ($\sim$0·1 eV), then the calculations are perhaps surprisingly good.

The calculated dipole moments of simple molecules also came out quite well from the theory, although Cook and McWeeny[1] found that for formaldehyde the CNDO method gave electron densities on the hydrogen atoms which were about 10% larger than those obtained from a non-empirical SCF calculation.

Wiberg[9] adopted the CNDO procedure (presumably CNDO/2), with some modifications to $U_{\mu\mu}$ and β^0_M, and calculated the heats of atomization of some hydrocarbons and their ions. This approach was moderately successful in fitting bond lengths and relative heats of atomization but the absolute values for the energies were very poor. Typical errors were 30 eV for methane, 41 eV for ethylene, 108 eV for benzene and 30 eV for acetylene. Wiberg applied a scaling factor to the molecular energies to bring them into better agreement with experiment. The resulting improvement is, however, at the expense of an additional empirical factor in the theory.

Sichel and Whitehead[10] with their parameters were able to improve on the binding energies obtained by Pople and Segal. In a study of the expressions for γ_{AB} they found that there was little to choose between the Mataga–Nishimoto (2.26) and Ohno–Klopman (2.28) expressions, but that both gave better results than those obtained using theoretical values. They evaluated β^0_{MN} from the average of β^0_{MH} and β^0_{NH}, where β^0_{MH} was chosen to fit the binding energy of a binary hydride at its equilibrium distance. Later Boyd and Whitehead[11] were able to further improve the binding energies by modifying the method by which β^0_{MN} was determined They chose to evaluate β^0_{MN} as the average of β^0_{MM} and β^0_{NN} where β^0_{MM} was obtained by fitting the experimental dissociation energy of the diatomic molecule M_2. The nuclear repulsion energy between the cores of atoms A and B was calculated from the expression

$$E_{CR}{}^{AB} = (P_{AA} - Z_A)(P_{BB} - Z_B)\gamma_{AB} \qquad (3.24)$$

which is similar to that used by Chung and Dewar[12] in their π-electron method. Although the calculated binding energies were in better agreement with experiment than the values obtained by Pople and Segal, the final results still left a great deal to be desired.

Fischer and Kollmar[13] in their CNDO calculations of the heats of atomization of hydrocarbons used an expression for the core-matrix elements $H_{\mu\mu}{}^c$ which was intermediate between CNDO/1 and /2 by taking†

† There is presumably a misprint in this expression in ref. 13 as the factor $(R_{AM}{}^2 + \zeta_\mu{}^2)$ appears with exponent $+\frac{1}{2}$. The boundary condition, $R_{AM} \to \infty$ would then be incorrect. We also presume that R_{AM} must be expressed in atomic units.

(see 2.16 and 3.12)

$$V_{A,\mu\mu} = Z_A[(1 - \alpha)\gamma_{AM} + \alpha(R_{AM}{}^2 + \zeta_\mu{}^2)^{-\frac{1}{2}}] \tag{3.25}$$

where ζ_μ is the valence orbital exponent and $\alpha = 0\cdot22$. If α is taken as zero the expression reduces to CNDO/2, with no penetration factor. If $\alpha = 1$ the expression approximates to the CNDO/1 procedure of calculating the integral (2.16) directly from Slater orbitals.

Fischer and Kollmar take $U_{\mu\mu}$ from ionization potentials as in CNDO/1, and $\beta_{\mu\nu}$ is given by a modified Wolfsberg–Helmholz formula (2.30) in which different weighting is given to the contributions $U_{\mu\mu}$ and $U_{\nu\nu}$. Although this procedure violates the condition for hybridization invariance the authors state that this is not an essential condition.[13a] The core repulsion was presumably taken as in CNDO/1 as the repulsion of nuclei with effective charges Z_A. With these parameters Fischer and Kollmar calculated the heats of atomization of the hydrocarbons shown in table 3.4.

TABLE 3.4   Heats of atomization of hydrocarbons calculated by a modified CNDO/1 method[13]

	Heat of atomization (eV)			Bond lengths (Å)		Force constants mdyne/Å	
	Calc.	Obs.	Type	Calc.	Obs.	Calc.	Obs.
Hydrogen	5·007	4·735	H—H	0·734	0·741	9·5	5·7
Methane	18·122	18·204	C—H	1·110	1·106	8·9	5·4
Ethane							
(staggered)	30·911	30·857	C—C	1·520	1·536	8·5	4·6
(eclipsed)	30·857						
Ethylene	24·353	24·408	C=C	1·342	1·332	14·7	10·9
Acetylene	17·769	17·605	C≡C	1·213	1·205	22·4	17·2
Propane	43·210	43·591	C—C	1·532	1·526		
Cyclopropane	36·925	38·313	C—C	1·513	1·524		
Allene	30·884	30·476	C=C	1·330	1·312	15·3	9·7
Butadiene	43·319	43·918	C=C	1·353	1·337		
Benzene	58·476	59·401	C—C	1·419	1·397	12·9	7·6

Inspection of results shown in table 3.4 reveals that although the overall figures are in quite good agreement with experiment, the method does give unsatisfactory results for the cyclic structures, benzene and cyclopropane both being predicted to be less stable than is determined experimentally. It is noteworthy that the method takes account of non-bonded interactions, as it correctly predicts the staggered form of ethane to be the preferred conformation. However the calculated barrier for internal rotation (1·5 kcal/mole) is 40% lower than the experimental values.

The method of Fischer and Kollmar represents the most successful attempt to date, to predict heats of atomization and equilibrium geometries within the CNDO method. Although these results might be improved further in a CNDO scheme, we feel that the improved results will still not achieve the desired chemical accuracy. This is because the CNDO method by its handling of electron repulsion through spherically arranged electron densities fails to introduce the full directional effects operating in a chemical bond. To obtain accurate molecular energies, we believe it will be necessary to use one of the more sophisticated methods yet to be described.

Attempts to predict electronic spectra with the original CNDO/2 method were rather unsuccessful,[14–16] the method giving excitation energies that were too high and with the excited states frequently in the wrong order. Kroto and Santry[17] have used the original CNDO/2 method to calculate electronic excitation energies and excited-state geometries. The excitation energy was calculated from the following expression

$$E_{\mu,v} = E_g + \epsilon_\mu - \epsilon_v - (J_{\mu v} - K_{\mu v}) \pm K_{\mu v} \qquad (3.26)$$

where E_g is the energy of the ground state with excited state geometry, ϵ_μ and ϵ_v the orbital energies (i.e. eigenvalues of the $\mathbf{F}$ matrix) and $J_{\mu v}$ and $K_{\mu v}$ the two-electron coulomb and exchange integrals respectively. The last term in the above expression is the correction that has to be made to the energy for singlet states (add $K_{\mu v}$) and triplet states (subtract $K_{\mu v}$). This method gave excited states in the correct order but the excitation energies were, in general, rather too high. This is perhaps not surprising for a method which minimizes only the ground-state energy, and one would expect better results if the excited-state energies were also minimized. Some excited-state geometries were given correctly by these calculations but others (e.g. the nonplanar excited state of formaldehyde) were incorrect.

In a later paper Kroto and Santry[18] describe an open-shell version of the CNDO/2 method which gave an improved agreement with the experimental results. For example, formaldehyde was then correctly predicted to be nonplanar in its lowest singlet excited state, with an out-of-plane equilibrium angle of $15°$; the observed value is in the range 20–40°.

Clark, in an attempt to improve the CNDO method for excited states, introduced the following modifications to the theory.[19] The integral γ_{AA} was calculated from the Pariser formula (2.20), and the integral γ_{AB} calculated from the Mataga–Nishimoto (2.26) expression. Sichel and Whitehead's values for $U_{\mu\mu}$ were adopted, and the core resonance integral $\beta_{\mu v}$ was calculated from the Wolfsberg–Helmholz expression (2.30)

Values for orbital exponents were taken from the compilation of Burns[20] on the grounds that the resultant orbitals were better approximations to the scf orbitals. This modified CNDO/2 method was applied by Clark to the compounds, cyclopropane, ethylene oxide and ethyleneimine; the results were however rather disappointing. In a further study Clark[21] compared the results obtained for pyrrole with *ab initio* results obtained by Clementi.[22] Orbital energies were in essentially the same order but the calculated charge distributions did not agree well.

At the present time probably the most successful CNDO treatment of excited states is that due to Del Bene and Jaffé.[23] They introduced three modifications into the CNDO/2 method with the aim of predicting reliable singlet–singlet excitation energies. The modifications were: (a) the evaluation of the one-centre integral γ_{AA} from the Pariser formula, (b) the use of the uniformly charged sphere model (2.23) to calculate γ_{AB}, and (c) the evaluation of $\beta_{\mu\nu}$ from the Wolfsberg–Helmholz expression (2.30) with the following values for k; for overlap between σ orbitals $k = 1$, and for overlap between π orbitals $k = 0.585$. Because of this difference in the values of k, the method can only be used for planar-conjugated molecules where the σ and π orbitals are orthogonal. With a configuration interaction limited to the thirty lowest singly excited states the method of Del Bene and Jaffé gave the singlet–singlet transitions

TABLE 3.5   Calculated $\pi–\pi^*$ singlet–singlet excitation energies and oscillator strengths by a modified CNDO method[23]

	Excitation energies (eV)		Oscillator strengths	
	Calc.	Obs.	Calc.	Obs.
Benzene	4·7	4·7	0·0	0·001
	5·2	6·1	0·0	0·100
	6·9	6·9	0·593	0·690
Pyrrole	5·0	5·7	0·080	
	5·4	6·5	0·006	
	7·0	7·1	0·129	
	7·0		0·479	
Furan	5·2	5·9	0·078	0·12
	5·8	6·5	0·009	0·08
	7·3	7·4	0·368	
	7·3		0·097	
Aniline	4·4	4·4	0·023	0·028
	4·7	5·4	0·041	0·144
	6·5	6·4	0·491	0·510
	6·6	6·9	0·434	0·570
Phenol	4·6	4·6	0·007	0·020
	5·0	5·8	0·010	0·132
	6·8	6·7	0·604	0·636
	6·8	6·9	0·537	0·467

shown in table 3.5.[23] The method has also been applied to some small molecules like formaldehyde and diazomethane.[24]

The results are generally satisfactory but no better than those obtained from a π-electron SCF method. It is probable, as the authors suggest, that the results could have been improved by evaluating γ_{AB} from the Mataga–Nishimoto relationship as this is the case for the Pople π-SCF method.

Vertical ionization potentials calculated by this method for some azabenzenes although in better agreement with the experimental values than those obtained from the original CNDO method were nevertheless still too high by approximately 1·5 eV.[25]

The effect of configuration interaction within the CNDO method has been studied by Giessner-Prettre and Pullman[26,27] using both the original parameters of Pople and Segal[3] and the modified scheme of Del Bene and Jaffé.[23] Transition energies calculated with the parameters of Pople and Segal were poor, even after an extensive CI treatment, but slightly better results were obtained with the parameters of Del Bene and Jaffé. The inclusion of doubly-excited configurations with the parameters of Del Bene and Jaffé led in most cases to a further improvement in the transition energies and in the calculated dipole moments.

Both the original CNDO/1 and CNDO/2 methods gave ionization potentials which were several electron volts higher than the experimental values, and they required an empirical correction ($\sim$4 eV) to bring them into agreement with the experimental values. More serious however was that the method predicted in many cases an incorrect ordering of orbital energies.

Sichel and Whitehead[28] have carried out a detailed study of ionization potentials within the CNDO method. They used their CNDO modifications already described, and studied the effect on the ionization potentials of varying the expressions for γ_{AB} and the values for β^0_M. They found that the semi-empirical values for γ_{AB} and β^0_M gave ionization potentials in better agreement with experiment than the corresponding theoretical values, and that there was little to choose between the Ohno–Klopman and Mataga–Nishimoto expressions. Table 3.6 shows some of the values obtained by Sichel and Whitehead. We note that there is sometimes an incorrect ordering of orbital energies (e.g. carbon dioxide) and this deficiency is independent of the parameter sets used.

Ionization potentials provide a test of the molecular eigenvalues, whilst dipole moments provide a test of the molecular eigenvectors. The theory behind the calculation of dipole moments within the CNDO approximation has been given by Pople and Segal.[3] They approximated the molecular dipole moment by a sum of two terms. The first gives the

TABLE 3.6 Comparison of observed and calculated vertical ionization potentials (eV)[28]

	Orbital symmetry	Calculated						Observed
		A	B	C	D	E	F	
Hydrogen	σ_g	14·69	15·44	18·62	17·11	18·60	20·09	15·45
Ammonia	a_1	12·45	13·30	15·47	13·26	14·00	16·30	10·35
	e	13·56	14·20	16·58	17·75	19·22	20·21	14·95
	a_1	27·07	26·71	30·50	34·59	35·89	37·08	
Water	b_1	14·11	14·33	17·90	14·11	14·72	17·83	12·61
	a_1	14·03	14·88	17·25	16·10	17·11	19·38	14·23
	b_2	14·68	15·36	17·85	18·64	20·22	21·44	18·02
	a_1	32·69	32·51	35·84	37·87	39·03	40·46	
Carbon dioxide	π_g	13·78	14·55	17·34	12·87	13·88	15·70	13·79
	π_μ	16·26	16·83	20·51	21·18	22·29	24·81	17·59
	σ_μ	13·58	14·55	16·38	17·24	18·23	20·43	18·07
	σ_g	20·69	21·01	23·26	21·03	21·81	24·42	19·36
	σ_μ	33·73	33·81	37·43	41·31	42·34	43·99	
	σ_g	34·67	34·46	38·18	43·09	44·06	45·50	
Methane	t_2	12·70	13·50	15·64	17·05	18·34	19·78	12·99
	a_1	23·51	23·40	26·16	32·32	33·56	34·68	~24·00

A. Using Mataga–Nishimoto expression for γ_{AB} (2.26). B. Using Ohno–Klopman expression for γ_{AB} (2.28). C. Using the theoretical γ_{AB}. D. Mataga–Nishimoto γ_{AB}, with Pople and Segal's β^0_A (with $\zeta_H = 1·2$). E. Ohno–Klopman γ_{AB}, with Pople and Segal's β^0_A. F. Theoretical γ_{AB}, with Pople and Segal's β^0_A.

contribution from the atomic charges:

$$\mu_{at} = 2·5416 \sum_A q_A R_A \tag{3.27}$$

where

$$q_A = Z_A - P_{AA}, \tag{3.28}$$

and R_A is the vector position of atom A. The second contribution arises from atomic dipoles resulting from the mixing of s and p orbitals of the same atom:

$$\mu_{sp} = -7·3370 \sum_A P_{sp}^{AA} \zeta_A^{-1}, \tag{3.29}$$

where ζ_A is the orbital exponent of atom A. Contributions from diatomic dipoles, that is the dipoles of overlap densities, were neglected. Thus the total molecular dipole moment is given by

$$\mu_{total} = \mu_{at} + \mu_{sp}. \tag{3.30}$$

The second term, μ_{sp}, does not occur in π-electron theory, and it may seem illogical to include such a term in the CNDO scheme when integrals which involve differential overlap between s and p orbitals are neglected.[†] The term is however known to be quite important, being largely responsible for the dipole moments, of molecules such as water and ammonia, and in many cases it is larger than μ_{at}. The results of Pople and Segal[3] listed in table 3.7 show this quite clearly.

TABLE 3.7 Dipole moments (Debyes) of some diatomic molecules calculated by the CNDO/1 method[3]. The negative end of the dipole is on the first atom written.

	μ_{at}	μ_{sp}	μ_{total}	μ_{exp}
LiH	−1·43	−4·59	−6·02	−5·90
BH	−0·61	2·55	1·94	—
CH	−0·32	1·87	1·55	—
NH	−0·09	1·42	1·33	—
OH	0·14	1·11	1·25	1·65
FH	0·36	0·91	1·27	1·91
LiF	−2·37	−1·78	−4·15	−6·60
BeO	−1·25	−3·18	−4·43	—
BN	−0·47	−2·80	−3·27	—
BF	0·24	2·46	2·70	—
CO	0·14	1·14	1·28	0·13

The fact that dipole moments are fairly well reproduced by the CNDO method has been substantiated by other workers with calculations on the fluorobenzenes,[29] other mono-substituted benzenes[30] and heterocyclic compounds.[31,32] Pople and Gordon[33] have calculated the dipole moments of a large number of organic molecules with a reasonable degree of success. Sichel and Whitehead,[34] in their detailed studies of the CNDO method, considered the effect on the dipole moment of varying γ_{AB}. They found that the dipole moment was rather sensitive to γ_{AB}, and furthermore that the Mataga–Nishimoto expression (2.26) gave the best results. Dipole moments calculated from expression (3.30) were in better agreement with the observed values, than dipole moments calculated using a point-charge model; this again underlines the significance of the atomic polarization term μ_{sp}.

Our attention so far has been confined to the first-row elements, which have a basis set composed of $2s$ and $2p$ atomic orbitals. One would expect difficulties to arise when we come to consider second-row elements because of the inclusion of $3s$, $3p$ and more especially $3d$ orbitals. However CNDO calculations have been performed with reasonable success on the

† The question of whether the basis orbitals in CNDO should be considered as orthogonal or non-orthogonal when calculating dipole moments has recently been considered.[59] The question is discussed further in section 5.9.

second-row elements, notably by Santry and Segal.[35] They considered the following three basis sets of atomic orbitals: sp, spd and spd'. In the first set (sp) all d orbitals are neglected so that the theory takes a similar form to that used for the first-row elements. The $1s$, $2s$ and $2p$ electrons are treated as part of the core. Penetration integrals were set equal to zero as in the CNDO/2 method, the integral $U_{\mu\mu}$ was evaluated from expression (3.19) with the quantities I_μ and A_μ being determined from atomic spectroscopic data. The coulomb repulsion integral γ_{AB} was evaluated theoretically, and $\beta_{\mu\nu}$ was calculated from the expression

$$\beta^0{}_{MN} = \tfrac{1}{2}k(\beta^0{}_M + \beta^0{}_N) \tag{3.31}$$

where $k = 0.75$ if either, or both, atoms M and N are second-row elements, otherwise $k = 1$. $\beta^0{}_M$ for the second-row elements were assumed to be proportional to those for the corresponding first-row element $\beta^0{}_{M'}$, according to the relationship

$$\beta^0{}_M = \beta^0{}_{M'} \left(\frac{U^M_{3s3s} + U^M_{3p3p}}{U^{M'}_{2s2s} + U^{M'}_{2p2p}} \right). \tag{3.32}$$

The second basis set adopted, spd, included $3d$ orbitals with the proviso that they have the same Slater exponent as the $3s$ and $3p$ orbitals. The theory is the same as for the first set except that the summation terms include $3d$ orbitals. In the third basis set taken, spd', the d orbital exponents were chosen so that the energy of an electron occupying a $3d$ orbital was approximately reproduced. The distinction made in this approximation between the $3d$ and $3s$, $3p$ orbitals will of course violate hybridization invariance. A further complication is that there are now three types of electron-repulsion integrals to be considered, $\gamma(s, s)$, $\gamma(s, d)$ and $\gamma(d, d)$, and two values for $U_{\mu\mu}$ and the core-attraction integrals $V_{A,\mu\mu}$. These are evaluated from the following expressions (cf. 3.19):

$$U_{\mu\mu}(s, p) = -\tfrac{1}{2}(I_\mu + A_\mu) - (Z_M - \tfrac{1}{2})\gamma_{MM}(s, s) \tag{3.33}$$

$$U_{\mu\mu}(d) = -\tfrac{1}{2}(I_\mu + A_\mu) - (Z_M - \tfrac{1}{2})\gamma_{MM}(d, s) \tag{3.34}$$

$$V_{A,\mu\mu}(s) = Z_A\gamma_{AM}(s, s) \tag{3.35}$$

$$V_{A,\mu\mu}(d) = Z_A\gamma_{AM}(d, s). \tag{3.36}$$

The γ integrals are all calculated from spherical Slater orbitals but with exponents equal to the s, p set or the d' set where appropriate. Santry and Segal adopted the same expression for $\beta_{\mu\nu}$ and the same values for β^0 as in the first two modifications.

In a study of dipole moments and bond angles they were able to ascertain the importance of $3d$ orbitals for these properties. They concluded that the inclusion of $3d$ orbitals was essential for accurate dipole moments, but that bond angles were less dependent on $3d$ orbitals being largely determined by the s and p overlap integrals. From an analysis of their results, they recommend that for calculations on second-row elements an spd' basis set be adopted with the following orbital exponents

$$\zeta_{3d}' = 0 \cdot 75 \zeta_{3p} \tag{3.37}$$

and with

$$\beta^0{}_{3d} = \left(\frac{2U_{3d3d}}{U_{3s3s} + U_{3p3p}} \right) \beta^0{}_{3s}. \tag{3.38}$$

In a later paper Santry[36] adopted the spd basis in a CNDO calculation of dipole moments and bond angles of second-row molecules. This treatment differed from the one described above, in the calculation of the resonance integrals, specific β^0 parameters being given for integrals involving d orbitals. For either μ or ν being a d orbital on atom M Santry proposed the relationship

$$\beta_{\mu\nu} = \beta_M{}^d S_{\mu\nu}. \tag{3.39}$$

For the case in which both μ and γ are d orbitals he took

$$\beta_{\mu\nu} = \tfrac{1}{2}(\beta_M{}^d + \beta_N{}^d) S_{\mu\nu}. \tag{3.40}$$

Some improvement in the results was noted when the **F**-matrix elements between orbitals of second-row elements on the same centre were reduced. Santry introduced an empirical factor of $1/4$ into (3.9) and took

$$F_{\mu\nu}{}^M = -\tfrac{1}{4} P_{\mu\nu} \gamma_{MM}. \tag{3.41}$$

The parameters determining the resonance integrals were deduced by comparing the CNDO calculations with some *ab initio* calculations using a gaussian basis set. The parameters U_{ss} and U_{pp} were also changed from the earlier work and U_{dd} was set equal to zero. With these modifications it was found that dipole moments were improved, but the calculated bond angles were slightly poorer.

Values for the valence-state ionization potentials and γ_{AA} of the second-row elements have also been determined by Sichel and Whitehead[5] and more recently by Levison and Perkins.[37] The latter work is more complete with respect to second-row elements, as these authors consider $3d$ orbitals.

This completes our description of the CNDO method, which as we have seen satisfactorily reproduces those properties dependent on the molecular eigenvectors (e.g. dipole moments), but is less successful for the prediction of properties dependent on the eigenvalues (e.g. molecular energies and ionization potentials).

3.5 Partial neglect of differential overlap, PNDO

In the CNDO method the invariance conditions require that all two-centre coulomb integrals $(\mu\mu \mid \nu\nu)$ be calculated in the s-orbital approximation. Examination of calculated values for these integrals suggests, however, that by adopting this approximation we may be losing a feature which has an important influence on molecular shape. For example, for p orbitals of carbon (with exponent 1·625) separated by 1·55 Å, the integrals have the values

$$(\sigma_a{}^2 \mid \sigma_b{}^2) = 10\cdot0 \text{ eV}$$
$$(\pi_a{}^2 \mid \pi_b{}^2) = 9\cdot0 \text{ eV}. \qquad (3.42)$$

The s orbital approximation would put both equal to 9·3 eV.

Dewar and Klopman[38] have argued that this dependence of the coulomb integrals on orbital shape is important in molecular calculations and they have developed a method which they call PNDO (partial neglect of differential overlap) which allows for its inclusion within a ZDO model.

The most important feature of the PNDO method is that integrals between two atoms M and N are evaluated only after transforming to a set of symmetry axes for the diatomic fragment MN. Thus the two-centre terms are calculated by a method which is independent of the choice of axes for the molecule as a whole. The initial basis set of atomic orbitals will usually be chosen to make use of any symmetry elements of the molecule; suppose this is a set φ_i. In setting up the SCF equations one will be faced with the evaluation of an integral like

$$(\varphi_\mu{}^M \varphi_\nu{}^M \mid \varphi_\rho{}^N \varphi_\sigma{}^N). \qquad (3.43)$$

The four orbitals involved can now be written in terms of orbitals s, px, py, pz, etc. defined with respect to diatomic axes as in figure 3.1. Most of the integrals that can be constructed from this diatomic set of atomic orbitals are zero by symmetry. For example $(s^M px^M \mid s^N s^N)$, and

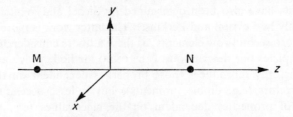

FIGURE 3.1   Diatomic axes in the PNDO method

$(s^M px^M \mid s^N pz^N)$ both refer to a situation in which one electron is on centre M with a function which is antisymmetric to reflection in the yz plane, but the other electron is on N with a function that is symmetric to this operation. The resulting repulsion of the two electrons must therefore be zero.

Dewar and Klopman divided the non-zero two-centre integrals into three classes:

1. $(s^2 \mid s^2)$, $(s^2 \mid px^2)$, $(px^2 \mid px^2)$, $(px^2 \mid py^2)$
2. $(pz^2 \mid px^2)$, $(pz^2 \mid s^2)$
3. $(pz^2 \mid pz^2)$ $\qquad$ (3.44)

These were evaluated from the following empirical expression

$$(\mu\mu \mid \nu\nu) = e^2(R_{\mu\nu}{}^2 + (\rho_\mu T_{\mu\nu} + \rho_\nu T_{\nu\mu})^2)^{-\frac{1}{2}}, \qquad (3.45)$$

the functions T depending on the class to which the integral belongs as follows:

Class 1,
$$T_{\mu\nu} = T_{\nu\mu} = 1.$$

Class 2,
$$T_{\mu\nu} = 1.$$
and
$$T_{\nu\mu} = e^{-R_{\mu\nu}/2(\rho_\mu + \rho_\nu)}. \qquad (3.46)$$

Class 3, both $T_{\mu\nu}$ and $T_{\nu\mu}$ are given by (3.46).

The scaled length ρ occurring in these expressions is determined from atomic spectral data by considering the united atom limit, $R_{\mu\nu} = 0$.

Expression (3.45) gives values of 7·13, 7·81 and 8·45 eV for integrals in classes 1, 2 and 3 respectively for a carbon–carbon bond length of 1·55 Å. Exact values of these integrals also separate roughly into three groups with approximately the same ratios.

The one-centre electron-repulsion integrals that arise in the PNDO method were obtained by an approximation proposed by Klopman.[39] Only two types of one-centre integral are, by symmetry, non-zero: these are of general form $(\mu\mu \mid \nu\nu)$ and $(\mu\nu \mid \mu\nu)$. Klopman assumed that the repulsion of two valence-shell electrons of an atom is the same whatever orbitals they occupy, but that their repulsion does depend on the electron spin. For electrons of the same spin the repulsion is labelled A^+, and for electrons of opposite spin A^-. Since the repulsion of electrons of opposite spin in orbitals μ and ν is just $(\mu^2 \mid \nu^2)$, we have

$$(s^2 \mid s^2) = (px^2 \mid px^2) = (px^2 \mid py^2) \text{ etc. } = A^-. \qquad (3.47)$$

However, for electrons of the same spin there is an exchange contribution to the energy which is therefore

$$(\mu^2 \mid \nu^2) - (\mu\nu \mid \mu\nu),\qquad(3.48)$$

we therefore have

$$(s\,px \mid s\,px) = (px\,py \mid px\,py) \text{ etc.} = A^- - A^+\qquad(3.49)$$

A^+ and A^- can be obtained from spectroscopic data together with the core energies U_{ss} and U_{pp}. ρ_μ, which arises in (3.45) and (3.46) can be determined from the one-centre limit ($R_{\mu\nu} = 0$) of (3.45) as follows:

$$\rho_\mu = e^2/2A_\mu^-.\qquad(3.50)$$

The Klopman approximation gives an average energy for singlet states of a configuration as the weighted average or 'barycentre' of all states of the configuration having $S_z = 0$. For the $2s^2 2p^2$ ground configuration of carbon for example

$$^1E = \tfrac{1}{9}[5E(^1D) + 3E(^3P) + E(^1S)].\qquad(3.51)$$

Note that this is not the average of all states as used in the CNDO method (3.14). An average triplet-state energy is also defined, which for carbon is just $E(^3P)$ as there is only one triplet state. In terms of A parameters we have

$$\begin{aligned}
^1E &= 2U_{ss} + 2U_{pp} + 2A^+ + 4A^-,\\
^3E &= 2U_{ss} + 2U_{pp} + 3A^+ + 3A^-.
\end{aligned}\qquad(3.52)$$

Similar expressions can be obtained for the energies of excited configurations and ions. Table 3.8 shows values of these parameters for three atoms.

TABLE 3.8   Klopman parameters for three atoms (eV)[39]

	$-U_{ss}$	$-U_{pp}$	A^+	A^-
C	49·884	42·696	10·144	11·144
N	69·593	58·669	10·718	11·975
O	96·247	80·591	12·149	13·707

The drawback of the Klopman method is that the one-centre contributions to the total energy are not invariant to rotation of the coordinate axes. We can show this by a simple example. Consider a diatomic system with two electrons occupying a molecular orbital

$$\psi = (pz^M + pz^N)/2.^{\frac{1}{2}}\qquad(3.53)$$

We have assumed zero overlap for normalization and taken the inter-nuclear axis as the z axis. The total electron-repulsion energy is

$$\tfrac{1}{4}((pz^{M} + pz^{N})^{2} \mid (pz^{M} + pz^{N})^{2}) \tag{3.54}$$

and the part of this which is made up of one-centre integrals of atom A is

$$(pz^{M}pz^{M} \mid pz^{N}pz^{N}) = A^{-}/4. \tag{3.55}$$

If we now redefine our coordinate axes such that the internuclear axis bisects the x and z axes, the molecular orbital (3.53) has the form

$$\psi = \tfrac{1}{2}(pz^{M} + px^{M} + pz^{N} + px^{N}) \tag{3.56}$$

and the electron-repulsion energy is

$$\tfrac{1}{16}((pz^{M} + px^{M} + pz^{N} + px^{N})^{2} \mid (pz^{M} + px^{M} + pz^{N} + px^{N})^{2}). \tag{3.57}$$

The part of this associated with atom M is

$$\tfrac{1}{16}[(pz^{M}pz^{M} \mid pz^{M}pz^{M}) + (px^{M}px^{M} \mid px^{M}px^{M}) + 2(pz^{M}pz^{M} \mid px^{M}px^{M})$$
$$+ 4(pz^{M}px^{M} \mid pz^{M}px^{M}) = \tfrac{1}{4}[2A^{-} - A^{+}]. \tag{3.58}$$

Expressions (3.55) and (3.58) are only equivalent if $A^{-} = A^{+}$, that is, if the integrals like $(pz\,px \mid pz\,px)$ are zero; in the Klopman method they are not.

This difficulty with the Klopman method only arises if there are bond orders $P_{\mu\nu}$ between different orbitals of the same atom. These are fre-quently zero, for reasons of symmetry, but if they are not then integrals like $(pz\,px \mid pz\,px)$ lead to a dependence of the energy on the coordinate axes. The defence of the Dewar–Klopman approach must rest on the fact that the lack of invariance usually leads to energies which vary only slightly with rotation of the axes, and that the results obtained by their method are in reasonable agreement with experiment. Before describing their results two more quantities need to be defined in the PNDO method.

The resonance integrals $\beta_{\mu\nu}$ were estimated from

$$\beta_{\mu\nu} = B_{MN}S_{\mu\nu}(I_{\mu} + I_{\nu})[R_{\mu\nu}^{2} + (\rho_{\mu} + \rho_{\nu})^{2}]^{-\frac{1}{2}} \tag{3.59}$$

where ρ is defined by (3.50). For bonds between different atoms the param-eters B_{MN} were chosen as the geometric mean of the parameters for bonds between like atoms B_{MM} and B_{NN} and these are to be determined by fitting the heats of atomization for selected compounds. For hydrocarbons the following were chosen: H_{2}, CH_{4}, $C_{2}H_{6}$, $C_{2}H_{4}$, $C_{3}H_{8}$.

The various parameters needed to build up the F-matrix elements have now been defined, but to obtain the total energies needed for heats of atomization or the calculation of equilibrium geometries, the core-repulsion energies are required.

In π-electron theory the core repulsion is usually set equal to the product of the core charges multiplied by the two-centre electron repulsion integral ($\gamma_{\mu\nu}$). If this relationship is used in all-valence-electron calculations however, the method leads to bond lengths which are considerably shorter than experimental values. The reason for its success in π-electron theory is that the sigma framework of the molecule gives the required rigidity and bond lengths are constrained to a narrow range; in the case of carbon between 1·30–1·52 Å. Dewar and Klopman[38] after trial and error adopted the following empirical expression for the core repulsion in the PNDO method

$$V_{MN} = \sum_{\mu(M)} \sum_{\nu(N)} \gamma_{\mu\nu} + \left[Z_M Z_N e^2 / R_{MN} - \sum_{\mu(M)} \sum_{\nu(N)} \gamma_{\mu\nu} \right] \exp(-\alpha_{MN} R_{MN}) \quad (3.60)$$

where the summation is over the valence orbitals μ and ν on atoms M and N respectively. Z_M and Z_N are the corresponding core charges and R_{MN} is the internuclear distance. The parameter α_{MN} was chosen to give the correct heats of formation of the standard compounds, and in order to reduce the number of parameters in the theory it was assumed that

$$\alpha_{MN} = (\alpha_{MM} \alpha_{NN})^{\frac{1}{2}}. \quad (3.61)$$

Expression (3.60) satisfies two essential boundary conditions. Firstly, at small internuclear distances, the expression approaches the point-charge model $Z_M Z_N e^2 / R_{MN}$. This is a harder potential than that obtained from the γ terms and so tends to prevent bonds becoming too short. Secondly, at large internuclear distances the net interaction between neutral atoms must be zero and for this to be so the core-repulsive forces must cancel out with the core-electron attraction forces. Expression (3.60) satisfies this condition, tending to γ_{MN} at large R_{MN}.

Unfortunately the aims of the PNDO method were not fulfilled as it failed to predict simultaneously molecular geometries and energies, and the heats of atomization shown in table 3.9 were obtained with assumed geometries.

Whilst the overall agreement between the calculated and observed values is good, it is noticeable that the method fails to give satisfactory heats of formation for those compounds with very short bond lengths such as allene and acetylene. Both of these compounds are predicted to be too stable by approximately 1 eV (20 kcal/mole) which suggests that the core repulsion as given by expression (3.60) does not rise steeply enough for short bond lengths. Nevertheless the PNDO method correctly distinguishes between

TABLE 3.9   Calculated heats of atomization and ionization potentials by the PNDO method (eV)[38,40,41]

	Obs.	Calc.	Obs.	Calc.
Ethane (staggered)	29·21	29·33	11·49	12·51
Ethane (eclipsed)	29·08	29·29	—	—
n-Butane	53·46	53·58	10·50	11·63
Isobutane	53·55	53·55	10·78	11·88
n-Pentane	65·56	65·71	—	—
Isopentane	65·64	65·66	—	—
Ethylene	23·28	23·39	10·48	10·86
Allene	29·24	30·19	—	—
1,3-butadiene (trans)	41·99	42·07	9·08	10·16
1,3-butadiene (cis)	41·89	42·06	—	—
Acetylene	16·96	17·93	11·36	11·06
Cyclohexane (chair)	72·74	72·76	9·79	11·51
Cyclohexane (boat)	72·66	72·52	—	—
Benzene	57·07	56·90	9·25	10·15
Cyclopropane	35·19	35·03	—	—

pairs of rotational isomers as shown by the results on ethane, cyclohexane and 1,3-butadiene. Conformational energies of the paraffin isomers were less satisfactory, as the branched isomers were predicted to be the stable species.

Baird and Dewar have confirmed the success of the PNDO method for cyclic compounds, with a series of calculations on conjugated[40] and strained cyclic[41] hydrocarbons. Although the accuracy achieved with the method (root-mean-squared deviation 2–3 kcal/mole) made it chemically useful, it did suffer from the restriction that experimental geometries had to be used which meant that the method could not be used to calculate reaction surfaces. Klopman[42] has used the PNDO method to calculate the possible structures of the norbornyl cation, but for the reasons just stated his results must carry little conviction. The PNDO method has since been superseded by a more satisfactory theory developed by Dewar's group which removes the need for using the experimental geometry; we will describe this method later.

3.6   Intermediate neglect of differential overlap—INDO

The approximation made in the CNDO method that all atomic states from a given configuration have the same energy is clearly a severe limitation as far as spectroscopy is concerned. One cannot, for example, obtain from the method any singlet–triplet splittings in the excited states.

Dixon[43] was the first to modify the CNDO method by retaining all one-centre exchange integrals like $(px\,py \mid px\,py)$. He called his method EMZDO (exchange modified zero differential overlap). Pople's group[44] made a similar modification and used the label INDO (intermediate neglect of differential overlap) and this is the label that is generally used.

We pointed out in the last section that with the exception of the coulomb and exchange integrals all the one-centre electron-repulsion integrals vanish through symmetry. In the INDO method both types of one-centre integral are retained, and the F-matrix elements have the following form (cf. 3.10 and 3.13)

$$F_{\mu\mu}{}^{\mathrm{M}} = U_{\mu\mu} + \sum_{\rho(\mathrm{M})} P_{\rho\rho}[(\mu^2 \mid \rho^2) - \tfrac{1}{2}(\mu\rho \mid \mu\rho)] + \sum_{\mathrm{A}}' (P_{\mathrm{AA}} - Z_{\mathrm{A}})\gamma_{\mathrm{MA}}$$
(3.62)

$$F_{\mu\nu}{}^{\mathrm{M}} = \tfrac{1}{2}P_{\mu\nu}[3(\mu\nu \mid \mu\nu) - (\mu\mu \mid \nu\nu)], \quad (\mu \text{ and } \nu \text{ both on M}) \quad (3.63)$$

and finally (cf. 3.9)

$$F_{\mu\nu} = \beta_{\mu\nu} - \tfrac{1}{2}P_{\mu\nu}\gamma_{\mathrm{MN}}, \quad (\mu \text{ on M and } \nu \text{ on N}). \quad (3.64)$$

The one-centre electron-repulsion integrals were expressed by Pople and coworkers[44] in terms of the so-called Slater–Condon parameters (F^n, G^n). These are electron-repulsion integrals after the angular co-ordinates have been integrated out. Typical relationships are the following:[45]

$$(s\,s \mid s\,s) = (px\,px \mid s\,s) = F^0 = \gamma_{\mathrm{MM}}$$
$$(s\,px \mid s\,px) = G^1/3$$
$$(px\,py \mid px\,py) = 3F^2/25 \qquad\qquad (3.65)$$
$$(px\,px \mid px\,px) = F^0 + 4F^2/25$$
$$(px\,px \mid py\,py) = F^0 - 2F^2/25$$

If we take as an example the configuration $2s^2 2p^2$ of carbon, then we can obtain F^2 from the separation of the ^1D and ^3P terms which is calculated to be $6F^2/25$, or the $^1S-^3P$ separation, $3F^2/25$. The mean value of F^2 obtained from these two values is $4\cdot7026$ eV. Slater[45] has obtained values for G^1 and F^2 for a number of elements from atomic spectroscopic data and some of these are shown in table 3.10.

The value for F^0 can be obtained in several different ways. Pople and coworkers[44] adopt the approach which they used in the CNDO method and calculate F^0 (i.e. γ_{MM}) analytically using Slater s orbitals. Baird and Dewar[46] follow a different procedure in which the one-centre terms U_{ss}, U_{pp} and F^0 are determined from experimental atomic transitions such as $s^n p^m \rightarrow s^n p^{m+1}$ and $s^n p^m \rightarrow s^n p^{m-1}$. Using theoretical values of G^1, F^2 and

TABLE 3.10 Values for the Slater–Condon parameters G^1 and F^2 (eV)[45]

	G^1	F^2
Li	2·4908	1·3498
Be	3·8087	2·4126
B	5·3941	3·5302
C	7·2469	4·7026
N	9·3670	5·9298
O	11·7546	7·2119
F	14·4095	8·5487

F^0 Pople and coworkers calculate the integrals $U_{\mu\mu}$ semi-empirically as in the CNDO/2 method from the mean of the ionization potential and electron affinity. The values obtained for $U_{\mu\mu}$ will of course differ from the corresponding values obtained in the CNDO/2 method because of the inclusion of the G^1 and F^2 terms. For example, the energy of an atomic state $2s^m 2p^n$ in the INDO approximation is given by

$$E = mU_{ss} + nU_{pp}$$
$$+ (m + n)(m + n - 1)F^0/2 - mnG^1/6 - n(n - 1)F^2/25 \qquad (3.66)$$

which can be compared with the CNDO expression (3.16).

The invariance conditions for the two-centre integrals (β and γ) must also be applied in the INDO method. Pople uses the same β^0_A as for the CNDO/2 method, and the γ_{AB} are likewise calculated analytically using Slater s orbitals.

Many molecular properties have been studied by the INDO method, particularly the spin densities of radicals which we shall describe in chapter 5. Table 3.11 compares results obtained from the CNDO and INDO

TABLE 3.11 Comparison of CNDO/2 and INDO calculations[44]

| | Equilibrium angle | | | Dipole moment (debyes) | | |
	CNDO	INDO	Obs.	CNDO	INDO	Obs.
$CH_2(^1A_1)$	108·6	107·2	102	2·26	2·17	—
$CH_2(^3B_1)$	141·4	132·4	136	0·75	0·53	—
OH_2	107·1	108·6	104·5	2·08	2·14	1·8
FH_2	180	180	—	0	0	—
CO_2	180	180	180	0	0	—
NO_2	137·7	138·5	132·0	0·75	0·79	0·4
O_3	114·0	115·5	116·8	1·26	1·09	0·58
NF_2	102·5	101·7	104·2	0·12	0·38	—
OF_2	99·2	99·0	103·8	0·21	0·40	0·3
NH_3	106·7	109·7	106·6	2·08	1·90	1·47
CF_3	113·5	111·6	111·1	0·17	0·68	—
NF_3	104·0	101·0	102·5	0·05	0·48	0·23

methods for the molecular geometries and dipole moments of some simple molecules.

These results show that the INDO method does not lead to any significant improvement in equilibrium bond angles or dipole moments. Both methods predict that $CH_2(^3B_1)$ is bent, contrary to an early analysis of the electronic spectrum but in agreement with a more recent one.[63] A more detailed study of equilibrium geometries by the INDO method has been made by Pople and Gordon.[47] Calculated bond lengths are rather better than those given by the CNDO method, particularly for X—H bonds. However, there are several poor results such as the O—O bond in H_2O_2 (calc. 1·22 Å, obs. 1·48 Å) and the C—C bond in C_2H_6 (calc. 1·46 Å, obs. 1·54 Å).

Relatively few ionization potentials have been reported with the INDO method. Table 3.12 shows a compilation of available data.[48–50]

TABLE 3.12   Calculated ionization potentials (eV) by the CNDO/2 and INDO methods[48–50]

	Orbital symmetry	CNDO	INDO	Obs.
1. Ammonia	a_1	12·3	10·2	10·8
	e	16·18	15·89	15·5
	a_1	33·0	33·55	
2. Carbon tetrafluoride	t_2	16·2	15·8	15·9
	t_1	18·2	17·3	17·2
	e	19·3	18·4	18·5
	t_2	24·3	23·1	(>21)
3. Borazine	$e'(\sigma)$	10·2	9·8	10·1
	$e''(\pi)$	11·0	10·4	11·4
	$a_1'(\sigma)$	14·6	14·0	12·8
	$e'(\sigma)$	14·9	14·8	14·0
	$a_2'(\sigma)$	16·7	16·6	14·7
	$a_2''(\pi)$	18·3	17·7	17·1

For all of these ionization potentials an empirical correction of minus 4 eV has been made to the eigenvalues of the F matrix. With this correction, the INDO method gives better ionization potentials than the CNDO method, and in particular gives better differences between successive ionization potentials.

Up to the present time few INDO calculations have been reported on excited states. Giessner-Prettre and Pullman[26] have studied the effect of CI in both the CNDO/2 and INDO methods, with particular reference to the molecules ethylene, formaldehyde, formic acid and formamide. Pople's original parameter set was taken in both methods and the excitation energies were generally at least 1 eV greater than the experimental values. However, these parameters were never intended for the calculation of electronic transitions and we have already seen that the CNDO method suitably parameterized (cf. the work of Del Bene and Jaffé) can give satisfactory excitation energies.

A large number of calculations on ground-state properties has been carried out by Dewar's group using a modified INDO method, and we will now describe their approach.

3.7 Modified intermediate neglect of differential overlap—MINDO

Baird and Dewar[46] introduced three modifications into the original INDO method with the specific aim of calculating heats of formation to chemical accuracy, that is, to one or two kcal/mole. There is a fundamental difference in philosophy between the Dewar and Pople schools. Pople has tried to reproduce the results of non-empirical SCF calculations using the same set of atomic orbitals, whilst Dewar is aiming to get the best fit to experimental data. Dewar would argue that by suitable parameterization one can, as in π-electron theory, get better results than non-empirical SCF calculations, because one can compensate for deficiencies in the basic theory that arise from an incomplete treatment of electron correlation. Thus in their modified INDO method, called MINDO/1, Baird and Dewar[46] choose their parameters to give the best fit to heats of formation of molecules in their ground states.

The one-centre integrals were evaluated as in the PNDO method which we described in section 3.5. The two-centre integrals γ_{MN} were evaluated from the Ohno–Klopman expression (2.28), with the value of γ_{MN} assumed to be dependent only on the type of atom and not on the type of orbital occupied (the invariance condition). The resonance integrals were calculated from the following expression:

$$\beta_{\mu\nu} = S_{\mu\nu}(I_\mu + I_\nu)(\beta_{MN}' + \beta''/R_{MN}^2) \tag{3.67}$$

where β'_{MN} and β''_{MN} were parameters chosen to fit heats of formation. Note that the valence-state ionization potentials appear in this expression and not the core integrals $U_{\mu\mu}$.

In contrast to the PNDO method, Baird and Dewar set the core repulsion, as in π-electron theory, to the product of the core charges multiplied by the two-centre electron-repulsion integrals. However, this expression leads to bond lengths that are considerably shorter than experimental values and much shorter even than those given by the PNDO method. Thus the method could not be used to calculate molecular geometries or reaction surfaces. Within the limitation of using assumed geometries, the MINDO method was used to calculate the heats of formation of hydrocarbons, some radicals, and some heteroatomic systems containing oxygen and nitrogen.[51] Calculated heats of formation were in extremely good agreement with the experimental values, the root-mean-square deviation between calculated and observed results being only 1·0 kcal/mole. There were, however, a few disturbing features of the method. It failed to give the correct relative energies of rotational isomers, for it predicted the following to be the more stable forms: *cis*-butadiene, axial-methyl-cyclohexane and *cis*-2-butene. Furthermore it predicted the staggered and eclipsed forms of ethane to be of almost equal energy. Thus although the MINDO method gave an accuracy in heat of formation greater than that achieved by any previous method, there was still plenty of room for improvement.

The MINDO method was subsequently improved by Dewar and Haselbach,[52] and in order to distinguish their method from that described above they called the new version MINDO/2. The principal modification was to calculate the core repulsion energy from the expression (3.60). The resonance integral β was evaluated from expression (3.67) but with β_{MN}'' set to zero. They were able, with suitable values for α (in 3·60) and β' (in 3·67) to obtain, simultaneously, heats of formation and molecular geometries with considerable accuracy. The foundation of their success was the use of a least-squares procedure to determine values for the parameters α and β'. Choosing the best values for these is not as easy as it might first appear because the parameters are interdependent and if one is changed, the optimum values of all the others are changed in an unpredictable manner. Thus an attempt to obtain the best α and β' parameters by trial and error is unlikely to succeed.

Dewar and Haselbach[52] obtained the following values for α and β' for a small set of hydrocarbons:

$$\beta'_{CC} = -0.3686, \quad \beta'_{CH} = -0.3410, \quad \beta'_{HH} = -0.4833 \text{ (eV)}$$
$$\alpha_{CC} = 1.6343, \quad \alpha_{CH} = 1.1843, \quad \alpha_{HH} = 0.6653, \text{ (Å)}^{-1}.$$

However, these values are just one of the many possible sets of near optimum parameters. In a later attempt[53] to find parameters that were more widely applicable α and β' were chosen by applying a least-squares procedure to the heats of formation and bond lengths of forty molecules containing the elements carbon, hydrogen, oxygen and nitrogen. Table 3.13 lists the values obtained.

It is interesting to compare these parameters with those previously obtained for hydrocarbons and to see that the H—H parameters have undergone the greatest change. There is perhaps a parallel here with *ab initio*

TABLE 3.13 Parameters for the MINDO/2 method[53]

	$\alpha(\text{Å})^{-1}$	$-\beta'(\text{eV})$
H—H	0·7535	0·35869
C—H	1·2475	0·33382
C—C	1·7345	0·35410
H—N	1·4204	0·37556
C—N	1·8551	0·36137
N—N	3·1459	0·29963
H—O	0·9073	0·54268
C—O	1·8657	0·43562
O—O	1·4547	0·68733
N—O	2·1550	0·42679

calculations where it is found that if orbital exponents are optimized, that of hydrogen is the most sensitive to molecular environments; optimum exponents for hydrogen can vary by as much as 35% from one compound to another.

Heats of formation, force constants, bond lengths and ionization potentials calculated with the parameters set of table 3.13 are shown in table 3.14. The results quoted include some of the compounds taken to define the parameterization but the results shown are typical of the accuracy that can be achieved with the MINDO/2 method. The results of table 3.14 have been obtained by varying one type of bond (indicated in parenthesis) with all others set equal to their experimental value. Two sets of results are quoted for ethane and nitromethane to show that the heat of formation is insensitive to small errors in bond length.

The overall results are certainly impressive, with an average error in heats of formation of 2·8 kcal/mole and in bond lengths of 0·013 Å. However, there are still some unsatisfactory features in the MINDO/2 method. The most important is that parameters which gave the best heats of formation gave all X—H bond lengths about 0·1 Å shorter than experimental values. As this shortening was consistent, Dewar and coworkers chose to make empirical corrections to all X—H bond lengths. These amount to +0·1 Å for C—H and N—H bonds and +0·15 Å for O—H bonds: the results shown in table 3.14 were obtained with this corrected geometry. This systematic error is of little chemical importance since it does not affect the overall molecular geometry.

The second unsatisfactory feature of the MINDO/2 method is that it gives relatively poor results for small cyclic compounds (in particular for three and four-membered rings), underestimating the strain energy in these compounds by approximately 10 kcal/mole.

TABLE 3.14 Results obtained by the MINDO/2 method[53]. Heats of formation with respect to elements in their standard states are given in kcal/mole. Force constants are quoted in mdyn/Å

Compound (bond)	Heat of formation		Bond length (Å)		Force constant		Ionization potential (eV)	
	Obs.	Calc.	Obs.	Calc.	Obs.	Calc.	Obs.	Calc.
$CH_4(C—H)$	−17·9	−16·2	1·093	1·100	5·0	5·8	12·70	13·16
$C_2H_6(C—C)$	−20·2	−21·7	1·543	1·512	4·5	4·9	11·56	11·46
$C_2H_6(C—H)$	−20·2	−22·2	1·093	1·109	4·8	5·7	11·56	11·46
$C_2H_4(C—H)$	12·5	14·4	1·337	1·317	9·6	9·3	10·51	10·58
$C_2H_2(C—C)$	54·3	58·0	1·058	1·061	5·9	6·1	11·40	10·90
$C_5H_{10}(C—C)$	−18·5	−26·5	1·534	1·530		5·5	10·49	10·42
$H_2O(O—H)$	−57·8	−58·9	0·957	0·972	7·8	10·1	12·62	12·20
$CH_3OH(O—H)$	−48·0	−51·8	0·960	0·973	7·6	9·5	10·83	10·96
$CH_3CHO(C=O)$	−39·7	−42·8	1·216	1·216		15·9	10·20	10·43
$HCOOH(C—O)$	−90·5	−89·9	1·312	1·338		11·3		11·21
$CH_3OOCH(Me—O)$	−81·0	−82·7	1·334	1·337		9·6		10·91
$CO_2(CO)$	−94·0	−92·9	1·162	1·179	16·8	22·7	13·68	12·92
$CH_3NH_2(C—N)$	−6·7	−9·0	1·474	1·435		5·7	9·18	9·86
$N_2H_4(NH)$	22·7	17·0	1·022	1·021		7·1		9·72
$NH_3(NH)$	−11·0	−11·2	0·912	0·918	6·4	7·7	10·16	10·48
$NH_4^+(NH)$	150·0	155·9	1·032	1·036	5·5	7·5		
$C_6H_5NH_2(CN)$	20·8	23·8	1·430	1·420		6·4	7·71	8·79
$HNO_3(HOON—O)$	−32·1	−36·4	1·22	1·208		15·8		12·81
$N_2O_4(NO)$	2·5	5·0	1·180	1·184		20·1		11·21
$CH_3NO_2(CN)$	−12·2	−14·8	1·47	1·485		6·0	11·23	11·59
$CH_3NO_2(NO)$	−12·2	−15·1	1·22	1·205		15·8	11·23	11·59

The equilibrium bond lengths and force constants in table 3.14 were all obtained by assuming a harmonic (parabolic) potential function and the ionization potentials were taken to be the negative of the orbital energy (Koopman's theorem). The force constants calculated with MINDO/2 are slightly too large but are considerably better than other SCF values (cf. table 3.4). This overestimation of force constants appears to be a general feature of all SCF methods, *ab initio* calculations included. Ionization potentials are in good agreement with experiment, and an improvement over those obtained with the CNDO and INDO methods.

Following this success of MINDO/2 it is natural that the method should be used to calculate quantities which are not directly determined from experiment, but which are needed to understand chemical phenomenon. Foremost in this area is a study of potential-energy surfaces for chemical reactions and in particular the determination of the energy and structure of transition states. Some calculations along these lines will be described in chapter 4.

3.8   Neglect of diatomic differential overlap—NDDO

The final method of calculation suggested by Pople, Santry and Segal in their important paper[2] is called the method of neglect of diatomic differential overlap (NDDO). Of the three levels of approximation which they considered it is the one that comes closest to the full SCF equations of Roothaan and is consequently more difficult to apply than the CNDO and INDO methods. Few NDDO calculations have been reported to the present time.

In all the ZDO methods an integral $(\mu\nu \mid \rho\sigma)$ is taken to be zero if μ and ν are orbitals of different atoms or if ρ and σ are orbitals of different atoms. The NDDO method differs from those described before in that all integrals $(\mu_A\nu_A \mid \rho_B\sigma_B)$ are retained in the calculation, and not just those for which $\mu = \nu$ and $\rho = \sigma$, as in the CNDO and INDO methods.

With this approximation added to those of the INDO method the F-matrix elements in the NDDO method are given by (c.f. 3.62–3.64)

$$F_{\mu\mu}{}^M = U_{\mu\mu} + \sum_{\rho(M)} P_{\rho\rho}[(\mu^2 \mid \rho^2) - \tfrac{1}{2}(\mu\rho \mid \mu\rho)]$$
$$+ \sum_{\rho,\sigma(A)}' P_{\rho\sigma}(\mu\mu \mid \rho\sigma) + \sum_A' V_{A,\mu\mu} \qquad (3.68)$$

$$F_{\mu\nu}{}^M = \tfrac{1}{2}P_{\mu\nu}[3(\mu\nu \mid \mu\nu) - (\mu^2 \mid \nu^2)] + \sum_{\rho,\sigma(A)}' P_{\rho\sigma}[(\mu\nu \mid \rho\sigma) - \tfrac{1}{2}(\mu\rho \mid \nu\sigma)]$$
$$+ \sum_A' V_{A,\mu\nu} \qquad (\mu, \nu \text{ both on atom M}) \qquad (3.69)$$

$$F_{\mu\nu} = \beta_{\mu\nu} - \tfrac{1}{2}\sum_{\rho(M)}\sum_{\sigma(N)} P_{\rho\sigma}(\mu\rho \mid \nu\sigma) \qquad (\mu \text{ on atom M}, \nu \text{ on atom N}).$$
$$(3.70)$$

Note that integrals $V_{A,\mu\nu}$, defined by (2.16) are not necessarily zero, also that the two-centre integrals are not given a common value depending on the type of atom.

The first detailed investigation of the NDDO method was made by Cook, Hollis and McWeeny.[54] In their treatment they evaluated all integrals theoretically using an orthogonalized atomic orbital basis. This is not therefore a method that should strictly be compared with the NDDO method as originally formulated. It is an approximate non-empirical SCF method rather than a semi-empirical method which directly uses experimental data in some part of the calculation.

Sustmann and coworkers[55] have described an NDDO method in which some of the integrals are given empirical values. The one-centre integrals $U_{\mu\mu}$ were taken from the compilation of Pople, Beveridge and Dobosh,[44] with the exception of that for hydrogen which was given the same value as in the MINDO procedure. The one-centre and two-centre electron-repulsion integrals were calculated analytically using Slater s orbitals. For hydrogen an orbital exponent of 1·0 was used, whereas the value usually adopted is 1·2. Penetration effects were ignored by setting the two-centre nuclear-electron attraction integrals $V_{A,\mu\nu}$ equal to the core charge multiplied by an electron-repulsion term as follows:

$$V_{A,\mu\nu} = -Z_A(\mu\nu \mid s^A s^A). \tag{3.71}$$

Finally, the core-resonance integral $\beta_{\mu\nu}$ was evaluated from the Wolfsberg–Helmholz expression (2.30). The empirical parameter k for hydrogen occurring in this expression was chosen to fit the energy of the hydrogen molecule at its equilibrium distance and values for the C—C and C—H bonds were obtained by fitting the bond lengths in methane and ethane. In a subsequent paper Allen and coworkers[56] modified the method slightly by fitting k_{HH} to give the correct bond length in hydrogen; accordingly k_{CC} and k_{CH} have different values in this work.

The principal application of the NDDO method by Sustmann and co-workers was to calculate the structures of some carbonium ions.[55] The ability of their method to predict experimental values was tested in a few cases. Some typical bond lengths calculated by their method are as follows: ethylene C=C 1·367 Å (observed 1·337 Å), acetylene C≡C 1·231 Å (observed 1·206 Å) and methane C—H 1·095 Å (observed 1·093 Å). The barrier to internal rotation in ethane was calculated to be 2·2 kcal/mole (experimental value 2·88 kcal/mole). The agreement between calculated and experimental data is quite good, but poorer than that obtained for MINDO/2.

The results of these NDDO calculations were at variance with *ab initio* calculations in that the bridged ions of protonated ethylene and acetylene were predicted to be more stable than their non-bridged isomers. This discrepancy was attributed to the neglect of some electron-repulsion integrals in the NDDO method which gives a tendency to overemphasize the importance of orbital overlap. It follows that those structures having the greatest number of bonds will have the lowest energies.

Allen and coworkers[56] have carried out a detailed comparison of the CNDO and NDDO methods. They adopted the CNDO method of Wiberg[9] for their calculations. Energy differences between conformational isomers, bond lengths, bond angles, ionization potentials, dipole moments and force constants were studied. Bond lengths calculated by the NDDO method were similar to those obtained by Sustmann and coworkers[55] and were slightly better than the CNDO values. There was little to choose between the bond angles predicted by the two methods, which were within 2° of the experimental values. Ionization potentials calculated with the NDDO method were, with the exception of that for H_2, in better agreement with experiment than the CNDO values, but in general the CNDO method gave better force constants although these were a factor of 2–3 larger than the experimental values. Heats of isomerization given by the two methods are compared with the experimental values in table 3.15.

TABLE 3.15   Heats of isomerization by the CNDO and NDDO methods (kcal/mole)

	CNDO	NDDO	Obs.
Ethane, staggered → eclipsed	2·32	2·39	2·9
Butadiene, *trans* → *cis*	0·63	−1·42	2·3
2-Butene, *trans* → *cis*	−0·25	−4·01	1·0
Butane, normal → iso	−0·19	−0·11	−2·0
Cyclohexane, chair → boat	5·02	4·83	6·6

The results are rather disappointing, since neither method gives the correct sign in all cases.

Allen and coworkers conclude from their work that, overall, the CNDO method gives better results than NDDO and deduce that the additional integrals which enter the NDDO calculation must have little effect on molecular energies. Cook, Hollis and McWeeny[54] from their comparative study are drawn to a different conclusion, for they find the CNDO method to be unreliable. It is our opinion, however, that any conclusions drawn from a comparative study where one, or both, of the methods does not use the optimum set of parameters, have little strength. Further refinement of the parameters in the NDDO approximation would probably lead to considerably better results.

At this point it does seem to be worth emphasizing that in a rather elaborate semi-empirical method like NDDO the determination of optimum parameters is not something that can be left to trial and error. The success of these theories in making predictions of chemical accuracy seems to be very dependent on the choice of parameters and these optimum parameters can only be determined by a technique similar to that adopted by Dewar and Haselbach.[52]

The other side of the picture, however, is that by careful parameterization one can certainly get a semi-empirical theory to be more successful than the non-empirical theory that it is derived from. Is one prepared to say that the semi-empirical theory therefore provides a better understanding of molecular properties? For example, a non-empirical SCF calculation on H_2O_2 with a minimum basis of atomic orbitals (hydrogen $1s$, oxygen $1s$, $2s$ and $2p$) does not give the correct structure of the molecule, which is a staggered configuration.[57,58] If a semi-empirical theory gave the correct structure with this basis, would one have any understanding of the factors influencing the molecular geometry from such a calculation? One might argue that any explanation based on these calculations would be false, and such a calculation could be used as an argument against careful parameterization to experimental data.

The CNDO and INDO methods have been reviewed in a book by Pople and Beveridge,[60] and Klopman and O'Lean[61] have reviewed the work principally covered by the Dewar group. Dewar's book[62] also describes some of the theories discussed in this chapter.

References

1. D. B. Cook and R. McWeeny, *Chem. Phys. Letters*, **1**, 588 (1968).
2. J. A. Pople, D. P. Santry and G. A. Segal, *J. Chem. Phys.*, **43**, S129 (1965).
3. J. A. Pople and G. A. Segal, *J. Chem. Phys.*, **43**, S136 (1965).
4. J. A. Pople and G. A. Segal, *J. Chem. Phys.*, **44**, 3289 (1966).
5. J. M. Sichel and M. A. Whitehead, *Theoret. Chim. Acta*, **7**, 32 (1967).
6. J. Hinze and H. H. Jaffé, *J. Amer. Chem. Soc.*, **84**, 540 (1962); *J. Phys. Chem.*, **67**, 1501 (1963).
6a. L. Oleari, L. Di Sipio and G. de Michelis, *Mol. Phys.*, **10**, 97 (1966).
7. R. L. Miller, P. G. Lykos and H. N. Schmeising, *J. Amer. Chem. Soc.*, **84**, 4623 (1962).
8. G. A. Segal, *J. Chem. Phys.*, **47**, 1876 (1967).
9. K. B. Wiberg, *J. Amer. Chem. Soc.*, **90**, 59 (1968).
10. J. M. Sichel and M. A. Whitehead, *Theoret. Chim. Acta*, **11**, 220 (1968).
11. R. J. Boyd and M. A. Whitehead, *J. Chem. Soc.* (A), 2598 (1969).
12. A. L. H. Chung and M. J. S. Dewar, *J. Chem. Phys.*, **42**, 756 (1965).
13. H. Fischer and H. Kollmar, *Theoret. Chim. Acta*, **13**, 213 (1969).
13a. H. Fischer and H. Kollmar, *Theoret. Chim Acta*, **12**, 344 (1968).
14. P. A. Clark and J. L. Ragle, *J. Chem. Phys.*, **48**, 4235 (1968) **46**, 4235 (1967).
15. P. A. Clark, *J. Chem. Phys.*, **48**, 4795 (1968).
16. P. M. Kuznesof and D. F. Shriver, *J. Amer. Chem. Soc.*, **90**, 1683 (1968).
17. H. W. Kroto and D. P. Santry, *J. Chem. Phys.*, **47**, 792 (1967).
18. H. W. Kroto and D. P. Santry, *J. Chem. Phys.*, **47**, 2736 (1967).
19. D. T. Clark, *Theoret. Chim. Acta*, **10**, 111 (1968).
20. G. Burns, *J. Chem. Phys.*, **41**, 1521 (1964).
21. D. T. Clark, *Tetrahedron*, **24**, 4689 (1968).
22. E. Clementi, H. Clementi and D. R. Davis, *J. Chem. Phys.*, **46**, 4725 (1967).
23. J. Del Bene and H. H. Jaffé, *J. Chem. Phys.*, **48**, 1807, 4050 (1968); **49**, 1221 (1968).
24. J. Del Bene and H. H. Jaffé, *J. Chem. Phys.*, **50**, 1126 (1969).
25. J. Del Bene and H. H. Jaffé, *J. Chem. Phys.*, **50**, 563 (1969).
26. C. Giessner-Prettre and A. Pullman, *Theoret. Chim. Acta*, **13**, 265 (1969).
27. C. Giessner-Prettre and A. Pullman, *Theoret. Chim. Acta*, **18**, 14 (1970).
28. J. A. Sichel and M. A. Whitehead, *Theoret. Chim. Acta*, **11**, 239 (1968).

29. D. W. Davies, *Mol. Phys.*, **13**, 465 (1967).

30. J. E. Bloor and D. L. Breen, *J. Phys. Chem.*, **72**, 716 (1968).

31. J. E. Bloor and D. L. Breen, *J. Amer. Chem. Soc.*, **89**, 6835 (1967).

31a. J. E. Bloor, B. R. Gilson and F. P. Billingsley, *Theoret. Chim. Acta*, **12**, 360 (1968).

32. N. S. Hush and J. R. Yandle, *Chem. Phys. Letters*, **1**, 493 (1967).

33. J. A. Pople and M. S. Gordon, *J. Amer. Chem. Soc.*, **89**, 4253 (1967).

34. J. M. Sichel and M. A. Whitehead, *Theoret. Chim. Acta*, **11**, 254 (1968).

35. D. P. Santry and G. A. Segal, *J. Chem. Phys.*, **47**, 158 (1967).

36. D. P. Santry, *J. Amer. Chem. Soc.*, **90**, 3309 (1968).

37. K. A. Levison and P. G. Perkins, *Theoret. Chim. Acta*, **14**, 206 (1969).

38. M. J. S. Dewar and G. Klopman, *J. Amer. Chem. Soc.*, **89**, 3089 (1967).

39. G. Klopman, *J. Amer. Chem. Soc.*, **86**, 1463 (1964).

40. N. C. Baird and M. J. S. Dewar, *Theoret. Chim. Acta*, **9**, 1 (1967).

41. N. C. Baird and M. J. S. Dewar, *J. Amer. Chem. Soc.*, **89**, 3966 (1967).

42. G. Klopman, *J. Amer. Chem. Soc.*, **91**, 89 (1969).

43. R. N. Dixon, *Mol. Phys.*, **12**, 83 (1967).

44. J. A. Pople, D. L. Beveridge and P. A. Dobosh, *J. Chem. Phys.*, **47**, 2026 (1967).

45. J. C. Slater, *Quantum Theory of Atomic Structure*, Vol. I, McGraw-Hill, New York, 1960.

46. N. C. Baird and M. J. S. Dewar, *J. Chem. Phys.*, **50**, 1262 (1969).

47. M. S. Gordon and J. A. Pople, *J. Chem. Phys.*, **49**, 4643 (1968).

48. G. R. Branton, D. C. Frost, F. G. Herring, C. A. McDowell and I. A. Stenhouse, *Chem. Phys. Letters*, **3**, 581 (1969).

49. D. C. Frost, F. G. Herring, C. A. McDowell, M. R. Mustafa and J. S. Sandhu, *Chem. Phys. Letters*, **2**, 663 (1968).

50. D. C. Frost, F. G. Herring, C. A. McDowell and I. A. Stenhouse, *Chem. Phys. Letters*, **5**, 291 (1970).

51. N. C. Baird, M. J. S. Dewar and R. Sustmann, *J. Chem. Phys.*, **50**, 1275 (1969).

52. M. J. S. Dewar and E. Haselbach, *J. Amer. Chem. Soc.*, **92**, 590 (1970).

53. N. Bodor, M. J. S. Dewar, A. Harget and E. Haselbach, *J. Amer. Chem. Soc.*, **92**, 3854 (1970).

54. D. B. Cook, P. C. Hollis and R. McWeeny, *Mol. Phys.*, **13**, 553 (1967).

55. R. Sustmann, J. E. Williams, M. J. S. Dewar, L. C. Allen and P. von R. Schleyer, *J. Amer. Chem. Soc.*, **91**, 5350 (1969).

56. R. B. Davidson, W. L. Jorgensen and L. C. Allen, *J. Amer. Chem. Soc.*, **92**, 749 (1970).

57. W. E. Palke and R. M. Pitzer, *J. Chem. Phys.*, **46**, 3948 (1967).

58. A. Veillard, *Theoret. Chim. Acta*, **18**, 21 (1970).

59. D. D. Shillady, F. P. Billingsley and J. E. Bloor, *Theoret. Chim. Acta*, **21**, 8 (1971).

60. J. A. Pople and D. L. Beveridge, *Approximate Molecular Orbital Theory*, McGraw-Hill, 1970.

61. G. Klopman and B. O'Leary, *Fortschr. Chem. Forschung*, **15**, 445 (1970).

62. M. J. S. Dewar, *The Molecular Orbital Theory of Organic Chemistry*, McGraw-Hill, 1969.

63. G. Herzberg and J. W. C. Johns, *J. Chem. Phys.*, **54**, 2276 (1971).

Chapter 4

Applications to chemical properties

In describing the theory of the SCF method in chapters 2 and 3, we concentrated our attention on those molecular properties which had been used to parameterize the SCF equations. These were, for the most part, heats of atomization, molecular geometries and electronic spectra. In this chapter we describe applications to other chemical properties, principally those of organic molecules, which have not been explicitly considered in the SCF parameterization.

4.1 Aromaticity

The term 'aromatic' was originally used to describe those cyclic conjugated compounds which displayed a chemical stability greater than that expected from the corresponding valence-bond structure. Experimentally, several criteria exist for defining the aromaticity, or otherwise, of a conjugated compound. Thus a compound may be termed 'aromatic' either by its having characteristic chemical properties, or by the existence of a ring-current as indicated by NMR spectroscopy. The degree of aromatic character exhibited by a compound has usually been measured by a quantity called the resonance energy.

Hückel[1] was the first to derive a set of rules from MO theory which predicted that monocyclic unsaturated systems would be aromatic if they contained $(4n + 2)\pi$ electrons, where n is an integer. Although this Hückel rule only strictly applies to monocyclic compounds, it has been applied by some authors to polycyclic systems; there is however, no justification for this from the derivation of the rule.

Hückel[1] attempted to put aromaticity on a more quantitative basis, when he defined resonance energy as the energy difference between a conjugated molecule containing $n\pi$ electrons and the appropriate number of

70

ethylene residues $(n/2)$ corresponding to a localized-double-bond structure. Thus the resonance energy of benzene was given by

$$E_R^{\pi} = E_{C_6H_6}^{\pi} - 3E_{C_2H_4}^{\pi} \tag{4.1}$$

Some typical values for the Hückel resonance energies are shown in table 4.1.

TABLE 4.1 Hückel resonance energies (units of β)

	Resonance energy
Benzene	2·00
Naphthalene	3·68
Anthracene	5·31
Azulene	3·36
Cyclooctatetraene	1·66
1,3-butadiene	0·47
Biphenylene	4·51
Pentalene	2·46
Heptalene	3·62

Clearly the HMO method is unsatisfactory for the prediction of chemical stabilities, since all the compounds listed in the table are predicted to have resonance-energy stability. The HMO method is in greatest error for the compounds pentalene and heptalene. Both are known to be highly unstable compounds, with the spectrum of heptalene clearly showing that it has an alternating (long–short) bond structure.[2]

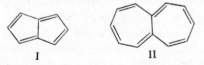

I   II

FIGURE 4.1   The structures of pentalene (I) and heptalene (II)

Another definition of resonance energy, which has found general acceptance, was given by Mulliken and Parr.[3] They defined resonance energy as the difference in energy between a conjugated compound and its corresponding classical structure: figure 4.2 shows diagrammatically the situation for benzene.

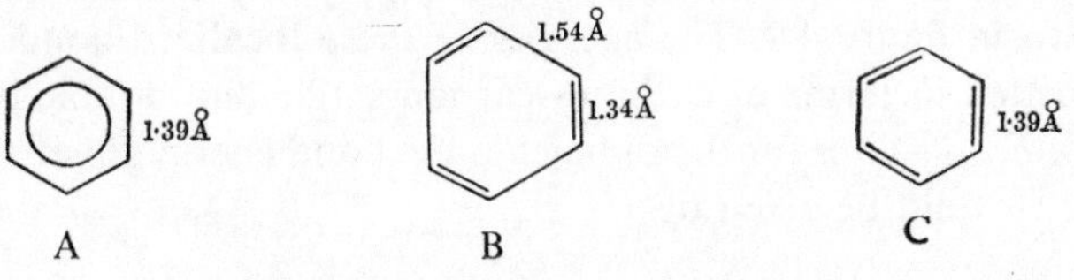

A   B   C

FIGURE 4.2   Valence structures for benzene

Thus the resonance energy of benzene would be given by the energy difference between benzene itself and 1,3,5-cyclohexatriene (B); this has been estimated to be 36 kcal/mole.[3] A vertical resonance energy has also been defined as the energy difference between benzene (structure A) and one of its Kekulé structures containing equal bond lengths, that is, structure C in figure 4.2. Various estimates exist for this vertical resonance energy of benzene, the most generally accepted figure being 63 kcal/mole.[4]

The Mulliken–Parr definition of resonance energy suffers from several weaknesses. Firstly, there is a certain ambiguity in the definition which is not readily apparent when we are discussing the relatively simple case of benzene. For more complex molecules several possible valence-bond structures may be drawn for the reference molecule, each of which may lead to a different value for the resonance energy. A second weakness is that resonance energy as defined by Mulliken and Parr, is not an experimentally determinable quantity, and even if it were, it is doubtful whether chemists would wish to know the stability of a compound relative to some hypothetical reference compound.

A third definition of resonance energy has been given by Dewar.[5] He defines it as the difference in heat of formation between a cyclic conjugated compound and its corresponding classical polyene. This definition hinges on the assumption that the π-electron delocalization energy of a polyene is very small, which is confirmed by π-SCF calculations. It appears from these calculations that the heats of formation and bond lengths of polyenes are close to the values expected for localized double bonds. Dewar defines compounds having a stability greater than the corresponding polyene to be aromatic and those which are less stable he calls antiaromatic. In the Dewar scheme resonance energy is defined to be positive for aromatic compounds and negative for antiaromatic compounds.

Lack of experimental data forced Dewar and Gleicher[6] to use their π-SCF method to calculate the energies and bond lengths appropriate for localized single and double bonds in the polyene. It would be expected that these values will conform very closely to the experimental values, since as we saw in chapter 2 the method is able to predict heats of atomization and bond lengths which are in good agreement with experiment.

Dewar and Gleicher[6] consider a series of polyenes and radialenes of the type shown in figure 4.3. The energies of these localized-bond structures can be written in terms of carbon—carbon single and double-bond energies E'_{C-C}, $E'_{C=C}$. Thus for the radialenes the bond energy E_C of the carbon framework would be given by

$$E_C = n(E'_{C-C} + E'_{C=C}). \tag{4.2}$$

E_C is then calculated by the π-SCF method, and is indeed found to be proportional to n. The analogous expression for the linear polyenes is

$$E_C = (n + 1)E'_{C=C} + nE'_{C-C} \qquad (4.3)$$

hence by combining data from both series it is possible to deduce values for both E'_{C-C} and $E'_{C=C}$. These are shown below[7] together with the values for isolated (as distinct from localized) bonds E_{C-C} and $E_{C=C}$ which were defined by the thermocycle (2.31) for sp^2–sp^2 bonds.

$$E'_{C=C} = 5\cdot54 \text{ eV}: \qquad E_{C=C} = 5\cdot56 \text{ eV}$$

$$E'_{C-C} = 4\cdot35 \text{ eV}: \qquad E_{C-C} = 3\cdot94 \text{ eV}.$$

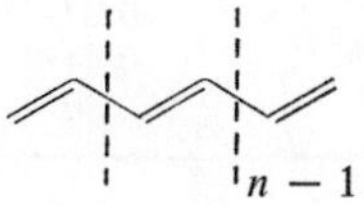

FIGURE 4.3 Structures of some polyenes (I) and radialenes (II)

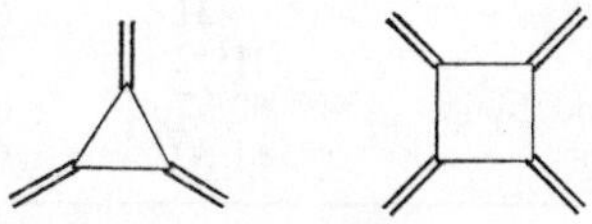

A significant feature of these values is the additional strength (approximately 0·4 eV) of the localized single bond compared with the isolated single bond. There is a resultant shortening from 1·51 Å (as taken by Dewar) to 1·46 Å. In contrast, the energy of the localized double bond is slightly less by 0·02 eV than the energy of a pure double bond, and has a bond length (1·35 Å) slightly longer than the length of a pure double bond (1·33 Å). Thus the calculations predict that the single bonds in a classical polyene will acquire some double-bond character. This in no way invalidates the condition for bond localization, it simply means that the single and double bonds in such systems can be thought of as localized, with the energy values $E'_{C=C}$ and E'_{C-C}.

It is now a simple matter to calculate the total energy of any classical polyene by addition of the individual $E'_{C=C}$ and E'_{C-C} terms, together with the total energy of the carbon—hydrogen bonds E_{C-H}. Thus in the Dewar scheme the resonance energy of any cyclic conjugated compound can be easily determined. If we again consider the particular case of benzene, then its resonance energy will be given by

$$E_R = \Delta H_a - (3E'_{C=C} + 3E'_{C-C} + 6E_{C-H}) \qquad (4.4)$$

where the heat of atomization ΔH_a could be the calculated or experimental value. At a later date, Dewar and coworkers[8] extended their

TABLE 4.2 Dewar resonance energies (eV)[7,8]

	Heat of atomization		Calculated resonance
	Obs.	Calc.	energy
Benzene	57·16	57·16	0·87
Naphthalene	90·61	90·61	1·32
Anthracene	123·93	123·89	1·60
Perylene	172·04	172·15	2·62
Triphenylene	157·76	157·94	2·65
1,3-butadiene	42·05	42·05	0·003
Azulene	89·19	89·46	0·17
Pentalene		70·53	0·006
Heptalene		108·15	0·09
Pyridine	51·79	51·87	0·91
Quinoline	85·18	85·32	1·43
Pyrrole	44·77	44·77	0·37
Aniline	64·31	64·34	0·86
Furan	41·52	41·56	0·07
Benzaldehyde	68·37	68·51	0·71
Phenol	61·60	61·67	0·89

definition of resonance energy, to include conjugated systems containing nitrogen and oxygen. Table 4.2 shows some of the results obtained.[7,8]

As we observed in chapter 2, and note again in the results of table 4.2, the heats of atomization calculated with the π-SCF method are in very good agreement with the experimental values. Furthermore the resonance energies calculated for the compounds seem to be in line with their chemical properties. Thus whereas the Hückel method had predicted substantial resonance energies for pentalene and heptalene, the above SCF results indicate that they have little aromaticity. Furthermore, although we have not included any structural results, both compounds are predicted to contain alternating single and double bonds of lengths approximately 1·46 and 1·35 Å. These results are therefore in agreement with the available chemical evidence.[2]

Furan, which is often considered to be aromatic, is estimated to have a relatively small resonance energy. Experimentally this seems to be the case since furan undergoes Diels–Alder reactions readily and a structural examination has revealed strong double-bond localization.[9] One further interesting feature of the above results, is that the resonance energy of triphenylene is three times that of benzene, and perylene has twice the resonance energy of naphthalene. Dewar has pointed out that this confirms the suggestion of Clar[10] that the central rings in such systems add little to the stability of the compound.

There are several advantages to Dewar's definition of resonance energy.

Firstly, there is no ambiguity about the structure of the reference compound, since it is clearly defined. Secondly, as we have shown, it is a relatively simple matter to calculate the resonance energy. Finally, although the resonance energies shown above were calculated, they could have been obtained directly from experimental data and this would be preferable. Baird has deduced the resonance energies of some hydrocarbons and carbonyl compounds with such a procedure.[11]

Lo and Whitehead[12] have also adopted an SCF approach to resonance energy, which they define as the difference in energy between a given cyclic conjugated compound and its most stable valence-bond structure. The energy of the compound was calculated by their π-SCF procedure, and the energy of the localized-bond structure calculated as a sum of the single and double-bond energies. The energy of the double bond was taken from that of ethylene, 5·60 eV, and the energy of the single bond was estimated to be 3·91 eV. The reader will note the basic similarity between this determination and that of Dewar. Lo and Whitehead calculate a term which they call the stabilization energy per carbon—carbon bond, which is simply the resonance energy divided by the number of carbon atoms in the compound. Values for the resonance energies obtained by Lo and Whitehead are considerably larger than the corresponding values obtained by Dewar mainly due to different values used for the energies of the single and double bonds.

One final point before we leave this section concerns the Hückel resonance energies shown in table 4.1. A comparison of these values with the corresponding SCF values shown in table 4.2, will reveal that they are in reasonable agreement if the Hückel values are corrected by $\frac{1}{2}\beta$ per C—C single bond, and β given the value of -40 kcal/mole. It is, however, extremely doubtful whether this correlation would hold for hetero-conjugated compounds in the face of the known unreliability of the Hückel method for such systems.

4.2 Chemical reactivity

The field of chemical reactivity provides the quantum chemist with a rich abundance of chemical data on which to test his theories. There is a vast literature on the subject based on the simple Hückel method which we will not attempt to review as it has been adequately described in several textbooks, some of which we referred to in chapter 1. In view of the evidence already presented in this book one must be sceptical of many of these results. Before we begin a discussion of the role of SCF methods in chemical reactivity we give a brief resumé of the basic equations of transition state theory.

In the transition state theory of chemical reactivity the rate of an irreversible reaction is determined by the difference in free energy between the reactants and the transition state. The standard equation for the rate constant is

$$K = \kappa \frac{kT}{h} \exp(-\Delta G^{\ddagger}/RT) \tag{4.5}$$

where κ is the transmission coefficient, usually taken as unity, and the other symbols have their usual meaning. The free energy of activation $\Delta G^{\ddagger}$ can be further subdivided into

$$\Delta G^{\ddagger} = \Delta H^{\ddagger} - T \Delta S^{\ddagger} \tag{4.6}$$

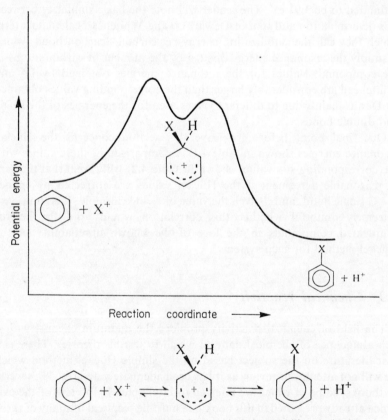

FIGURE 4.4   Free energy profile for a typical electrophilic substitution

where $\Delta H^{\ddagger}$ and $\Delta S^{\ddagger}$ are respectively the enthalpies and entropies of activation.

It is not possible at the present time to calculate the absolute rate of a chemical reaction from formula (4.5) because to determine $\Delta H^{\ddagger}$ and $\Delta S^{\ddagger}$ we would need a complete knowledge of the potential energy surface of the reactants. Thus we are forced to calculate the relative rates of reactions, making the assumption that changes in entropy, and the effects of solvation, will be constant for a set of similar reactions.

Let us consider as a typical reaction the electrophilic aromatic substitution illustrated in figure 4.4. It is likely that the free energy profile of such a reaction would be similar to that shown.

The intermediate in the above reaction has been called a σ complex or Wheland intermediate.[13] This may or may not have a well-defined structure, and is generally considered as a loose addition complex in which the group X (NO_2^+, H^+, etc.) and the leaving hydrogen atom are on opposite sides of the benzene plane with the attacked carbon atom in an approximately tetrahedral configuration. The σ complex is not the transition state for the reaction, as shown in the figure, but is usually assumed to be close to the transition state on the potential-energy surface: how close depends upon the type and conditions of a reaction.

Several different reactivity indices have been invoked in Hückel theory to correlate with chemical reactivity. The indices have been classified into two categories according to whether they referred to the initial stages of a chemical reaction or to the transition state. The indices in the first category are collectively associated with the isolated-molecule method. Since we will not refer to these indices in detail, we shall not define them here but simply state that they are: charge density, free valence, self-polarizability, frontier electron density and super-delocalizability. The second category is composed of the localization energy index which is based on the energy required to form the Wheland intermediate. For aromatic substitution this index can be divided into three types L_{μ}^+, L_{μ}, L_{μ}^-, corresponding to the three types of reactions, electrophilic, radical and nucleophilic. It is defined, and calculated, as the energy required to localize 2, 1 or 0 π electrons respectively at some position μ in a conjugated molecule; that position being subsequently removed from the rest of the conjugated system.

Whilst the isolated molecule method was reasonably successful in correlating chemical reactivities, discrepancies did arise. A well known example of this is the case of electrophilic substitution in the fluoranthene molecule (figure 4.5), which experimentally[14] is found to occur in the order

$$3 > 8 > 7 > 1 > 2.$$

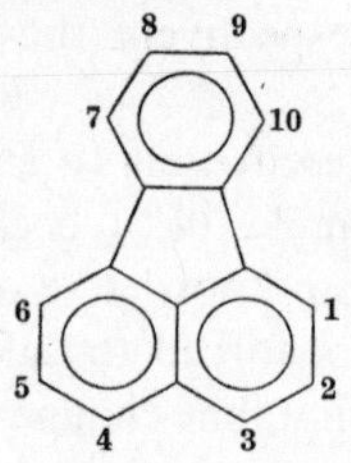

FIGURE 4.5   The structure of fluoranthene

Values obtained from Hückel theory for the various reactivity indices are shown in table 4.3

TABLE 4.3   Hückel reactivity indices calculated for fluoranthene

Position	Charge density	Self-polarizability	Frontier electron density	Super-delocalizability	Localization energy
1	0·947	0·440	0·453	0·828	2·466
2	1·005	0·400	0·398	0·860	2·503
3	0·959	0·462	0·470	0·903	2·341
7	0·997	0·427	0·438	0·936	2·371
8	1·008	0·510	0·409	0·872	2·435

Inspection of the results shows that all the indices fail to predict the observed order of reactivity for all positions. The localization energy clearly gives the best agreement, the only anomaly being the order predicted for the 7 and 8 positions. There are other examples which show that the localization energy index is, in general, more reliable than any of the isolated molecule indices.

Chalvet, Daudel and Kaufman[15] have carried out a detailed study of the localization energy obtained by the Hückel, Pariser-Parr and the Pople π-SCF method. There was general agreement between the three methods in predicting the most reactive position to chemical attack. However, correlation with experimental results revealed that the Pople π-SCF method gave better agreement than the simple Hückel method.

This work was criticized by Dewar and Thompson[16] on the grounds that the core-repulsion energy had been neglected in the calculation of the energy terms. Thus the localization energy calculated by Chalvet, Daudel and Kaufman referred to the difference in π-electron energies E_π, and not the required π-bonding energy E_{π_b} which includes the core-repulsion energy E_{CR}

$$E_{\pi_b} = E_\pi + E_{CR}. \tag{4.7}$$

If the difference between E_{CR} for the reactant and intermediate were a constant then both E_{π_b} and E_π would predict the same result. In practice, however, this is not so, since the value of E_{CR} for the transition state depends upon the position of substitution. Inclusion of the term E_{CR} by Dewar and Thompson in their SCF study of chemical reactivity, led to an improved agreement over that obtained by Chalvet, Daudel and Kaufman.

TABLE 4.4 Hückel and SCF localization energies for some hydrocarbons relative to benzene

	Position of substitution	Hückel L_μ^+ $(-\beta)$	SCF L_μ^+ (eV)
Benzene		0·000	0·000
Naphthalene	1	−0·237	−0·778
	2	−0·056	−0·566
Anthracene	1	−0·286	−1·186
	2	−0·136	−0·941
	9	−0·523	−1·585
Pyrene	1	−0·346	−1·465
Perylene	3	−0·396	−1·717
Chrysene	6	−0·285	−1·273
Coronene		−0·230	−1·379
Triphenylene	1	−0·158	−0·908
	2	−0·059	−0·935
Biphenyl	2	−0·136	−0·689
	4	−0·089	−0·813

Electrophilic, nucleophilic and radical localization energies were calculated by Dewar and Thompson and correlated against the corresponding chemical reactions for a series of conjugated hydrocarbons.[16] A comparison with the corresponding values from Hückel theory revealed the superiority of the SCF values. Table 4.4 shows some of their electrophilic localization energies calculated relative to benzene; thus a negative value indicates a chemical reactivity greater than that of benzene.

Dewar and Thompson showed that for both the rate of nitration and the basicity (the equilibrium constant for proton addition being the relevant quantity) the SCF localization energies gave a better correlation with the experimental data than the Hückel parameters. Their results are shown in figure 4.6.

As we have already stated, the localization energy is calculated on the basis that the Wheland intermediate is a close approximation to the transition state; this will not always be the case, and deviations from it will be reflected in the slope of the correlation line. For an 'early' transition state, (i.e. one closer to the reactant side of figure 4.4) the hybridization at the carbon atom under attack will be closer to sp^2 than to sp^3 and therefore conjugation of this atom in the transition state will still be significant. Hence the actual localization energy will be much smaller than the calculated value, with a consequent reduction in the slope of the correlation line. The limiting case would be an attacking reagent of such high reactivity that the transition state would resemble the isolated molecule. In this

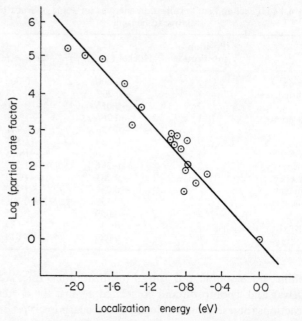

FIGURE 4.6   Relative electrophilic localization energies plotted against the logarithms of partial rate factors for nitration[16]

respect the slope of the correlation line resembles the Hammett reaction constant ρ.[17] Unfortunately quantitative results for the slopes of the SCF localization energies against relative rates are lacking.

The fact that localization energies calculated by the Pople π-SCF method are in quite good agreement with the relative rates of a chemical reaction would seem to confirm our assumptions of constant entropy and $\sigma-\pi$ separability. However, from experiment it is well-known that entropy is not constant even for a similar series of compounds[17] and the success of the theory appears to be due to the fact that for reactions in solution $\Delta H^{\ddagger}$ and $\Delta S^{\ddagger}$ show parallel changes within such a series. Qualitatively one can understand this result by considering the effect on the free energy of

activation of making a small structural change λ in some component.†
This can be expressed mathematically by differentiating expression (4.6)
with respect to λ, that is,

$$\frac{\partial \Delta G^{\ddagger}}{\partial \lambda} = \frac{\partial \Delta H^{\ddagger}}{\partial \lambda} - \frac{T \partial \Delta S^{\ddagger}}{\partial \lambda} . \tag{4.8}$$

We know that the free energy of activation will always be at a minimum
with respect to any structural change or 'perturbation' of the reaction.
This is not however the case for $\Delta H^{\ddagger}$ and $\Delta S^{\ddagger}$, thus it follows that the
above expression reduces to

$$\frac{\partial \Delta H^{\ddagger}}{\partial \lambda} = \frac{T \partial \Delta S^{\ddagger}}{\partial \lambda} , \tag{4.9}$$

or in other words, $\Delta H^{\ddagger}$ and $\Delta S^{\ddagger}$ are mutually compensating. This argument, which is given by Streitwieser,[14] applies to the perturbation of any
one molecule. There are however no theoretical reasons why $\Delta H^{\ddagger}$ and
$\Delta S^{\ddagger}$ should be exactly correlated within a series of molecules.

Streitwieser and Langworthy[19] have studied the rates of deprotonation
of some arylmethyl derivatives which proceed according to the reaction

$$ArCH_3 + B^- \rightarrow ArCH_2^- + HB.$$

In a correlation of their experimental rate data with the π-bonding energy
E_{π} calculated with the Hückel method, they found that the results divided
into two groups of α and β naphthyl types. They attributed this dichotomy
to the greater steric interaction present in the α-naphthyl compounds,
that is, to the repulsion between the peri hydrogen and the methylene group.
This effect has also been observed by other workers.[14,20,21] However,
Dewar and Thompson[16] in their SCF study found that all the experimental
data could be represented by one correlation line. They considered the
Hückel results to be in error, and the occurrence of the two correlation
lines they attributed to the known deficiencies of the Hückel method.
Altschuler and Berliner[22] measured the rates of bromination of some aromatic hydrocarbons and found a reasonable correlation with the SCF
localization energies of Dewar and Thompson.[16] However, by comparing
the rates of bromination with the corresponding rates of solvolysis they
found that although there was considerable scatter most of the points did

† We are using here the concept of a free-energy surface, which is not wholly
accepted as part of transition-state theory, but which has been proposed by
M. Schwartz[18] and others.

fall into two groups of α and β naphthyl types, which, the authors suggested, confirmed the Hückel results. Gleicher[23] modified the Hückel procedure for some α and β arylmethyl compounds by adding non-bonded interactions calculated empirically. The modified Hückel results gave an even poorer correlation with experiment than had been obtained previously, whereas the same modification with the SCF method led to an improved correlation.

The π-SCF methods have also been applied to the study of equilibrium constants. The equilibrium constant of a reversible reaction is a measure of the free-energy difference ΔG between the reactants and the products, that is,

$$K = \exp(-\Delta G/RT). \qquad (4.10)$$

The assumptions of constant entropy and σ–π separability that were made earlier, are again adopted.

Typical equilibrium reactions have been treated by Kende[24] who considered the basicity of some aldehydes and ketones, a reaction that can be summarized as follows:

$$R_2CO + H^+ \rightleftharpoons R_2\overset{+}{C}OH.$$

The π-electron energy required for protonation was calculated as the difference in energy between the neutral carbon species and the corresponding protonated species. Results calculated with the Hückel method were rather poor, but those calculated with the SCF procedure gave a good correlation with pK_a's (correlation coefficient 0·90).

Dewar and Morita[25] in their SCF study of the basicity of carbonyl compounds were able to improve on the correlation obtained by Kende. They attributed the improvement to three factors. Firstly, their SCF results were obtained from a variable β procedure, whilst Kende had carried out a fixed β procedure with all C—C bonds 1·40 Å, and bond lengths of 1·23 Å for C=O and 1·38 Å for C$\overset{+}{=}$OH. Secondly, Kende had taken the same geometry for the carbon framework of both the protonated and unprotonated species. The third criticism was the same as that used against the work of Chalvet, Daudel and Kaufman,[15] namely that the localization energy had been calculated on the basis of π-bonding energies and not from the total energy. The above criticisms would also apply to SCF calculations which have been made of the basicity of some methylbenzenes[26] and some nitrogen heterocyclic molecules.[27]

A more rigorous method of determining free-energy differences, would be to use one of the all-valence-electron SCF–MO methods described in chapter 3, since these take explicit account of the σ electrons. At the present time few all-valence-electron calculations have been performed on chemical reactivity, and some of these have not been very encouraging.

Lewis[28] has used the CNDO/2 procedure to study the acidities of alkanes and saturated alcohols and the basicity of amines. The energy difference between the neutral and ionic species was calculated using an assumed geometry for both species, and this difference in energy was equated to the corresponding equilibrium constant. The order of acidity predicted for the alkanes was in complete disagreement with experiment, although, as the author suggests, this may have been due to the fact that the experimental values referred to solution. The order of acidity for the alcohols and the order of basicity for the amines was, with one exception in the alcohol series, in agreement with the experimental order.

Jesaitis and Streitwieser[29] have also carried out a CNDO/2 study on the acidity of the cycloalkanes. Assumed geometries were used for the hydrocarbons and the ionic species. The results obtained were in poor accord with experiment as they predicted an order in reverse to that found experimentally. This led Jesaitis and Streitwieser to formulate a new method called IRDO (intermediate retention of differential overlap), which basically was an NDDO treatment for bonded atoms but which reduced to a CNDO procedure for non-bonded atoms. Even with the inclusion of these extra integrals the results obtained, although better than with CNDO/2, were still in poor agreement with experiment.

Baird[30] has used MINDO/1 to calculate the acidity of some saturated alcohols. The order predicted with this procedure was in agreement with that observed in the gas phase.

The results obtained so far for aromatic substitution are rather better. Streitweiser and coworkers have used the CNDO/2 method to calculate L_μ^+ values for aromatic hydrocarbons and some of their methyl derivatives.[31] The results were found to be much better than those obtained with the Hückel method when tested against the rate of hydrogen–tritium exchange in acid media. However, the correlation line showing the influence of methyl substitution had a very different slope from that found for the family of aromatic hydrocarbons. Also the results for benzocyclobutene and fluorene were poor. There is therefore still a great deal of room for improvement.

The above all-valence-electron calculations have used assumed geometries throughout. With the availability now of new methods (e.g. MINDO/2) which give simultaneously satisfactory molecular energies and geometries, the restriction of taking assumed geometries is removed, and so a considerable improvement in results can be expected. Confusing as the present picture may seem, it still gives encouragement for future work in this field, and confidence that the all-valence-electron methods will make a substantial contribution to the subject of chemical reactivity.

4.3   The hydrogen bond

The hydrogen bond has been of interest to chemists for many years but it has recently received increased attention because of its importance in biological systems.

A hydrogen bond represents a situation in which a hydrogen atom appears to be bonded to two atoms as suggested by the structure

$$A—H \cdots B.$$

In most cases the components A—H and B, which may be in the same molecule or in different molecules, preserve their identities on association so that one has only a slightly perturbed AH + B situation. In other cases, notably the ion $(FHF)^-$, the hydrogen is shared equally by the two atoms and the hydrogen bond is much stronger. This situation, which is somewhat analogous to the sharing of hydrogens in the boron hydrides, is perhaps better considered outside the typical hydrogen-bond classification.

We will not attempt to review the early attempts to calculate the hydrogen-bond energy by the electrostatic model, or by a valence-bond approach, or by the more recent general theory of intermolecular forces. Some aspects of this work have been reviewed.[32,33] In this section we describe only the empirical SCF calculations.

Although hydrogen-bonding has been extensively studied with the SCF methods, most of the calculations reported have been based on the π-SCF theory with appropriate modifications to the core parameters to represent the hydrogen-bonding. Although the first two topics described in this chapter lend themselves fairly readily to treatment by π-electron theory, the phenomenon of hydrogen-bonding can only be satisfactorily treated if the σ electrons are explicitly considered, because the energy of the A—H $\cdots$ B σ bond cannot be considered as a constant even for a related series of molecules. We shall therefore exclude from our discussion those studies on hydrogen-bonding which have used the π-SCF theory. We refer the reader interested in this work to the review by Bratoz.[33]

(a) Water

Water represents one of the simplest examples of an asymmetric hydrogen bond. The hydrogen bonding between water molecules has been treated

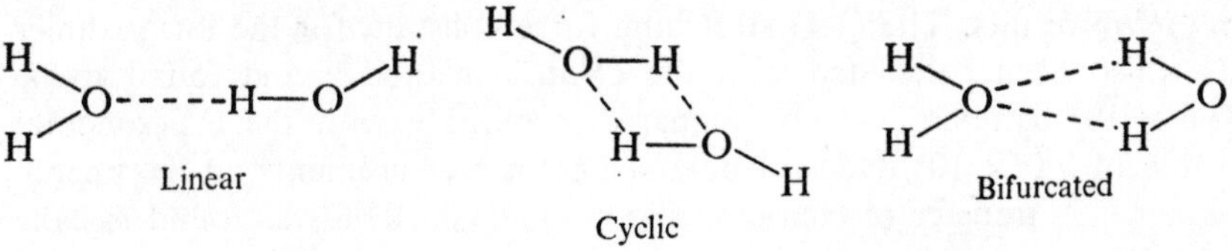

FIGURE 4.7   Possible structures for the water dimer

with the CNDO/2 and NDDO methods, and *ab initio* calculations are also available.

Hoyland and Kier[34] have used the CNDO/2 method to calculate the water-dimer energies, and have considered the three structures shown in figure 4.7. Kollman and Allen[35] have compared the CNDO, NDDO and *ab initio* results on these structures, and Morokuma and Pedersen[36] have also made *ab initio* calculations. Murthy and Rao[37] in their CNDO/2 study considered only the linear conformation of the water dimer. All the CNDO/2 methods used the parameter set of Pople and Segal.[38]

The procedure adopted by the above authors was to calculate the total energy of the dimer as a function of the O–O distance. The geometry taken for the monomer unit within the dimer was not varied, it being assumed to have either the calculated or the experimental equilibrium geometry of water. Calculated energies of stabilization obtained with this procedure, together with the O–O distances are shown in table 4.5.

The CNDO and *ab initio* methods predict the same order of stability for the three conformations, namely linear > bifurcated > cyclic. Recent experimental evidence is in favour of the linear structure being the most stable.[39] Although the absolute energy values calculated by the CNDO and *ab initio* methods are in poor agreement, there is slightly better agreement between their energy differences. The O–O distance is estimated experimentally to be in the range 2·74–2·77 Å,[40] which is close to the

TABLE 4.5   Stabilization energies and O–O distances for some possible structures of the water dimer (energies in kcal/mole, bond lengths in Å)

| Linear | | Cyclic | | Bifurcated | | Type of |
Energy	O–O	Energy	O–O	Energy	O–O	calculation
12.6	2·6	7·8	2·75	9·2	3·0	*Ab initio*[36]
5·3	3·01	4·0	2·90	4·45	3·0	*Ab initio*[35]
6·31	2·54	1·91	2·67	2·84	2·41	CNDO[34]
5·94	2·53	2·31	2·25	2·50	2·44	CNDO[35]
6·27	2·54					CNDO[37]
76	2·2	76	1·6	60	2·0	NDDO[35]

ab initio results. The O–O stretching force constant for the linear dimer has also been calculated with the CNDO/2 method,[37] and found to be 7.6×10^4 dyne/cm, which compares favourably with the experimental value of 7.0×10^4 dyne/cm obtained from measurements on the phenol dimer.[41] A transfer of charge in the range 0·006–0·06 was found to take place between the donor and acceptor water molecules on dimerization.[34–37]

The higher polymers of water have been studied by Hoyland and Kier,[34] who considered two linear conformations for the water trimer and one conformation for a pentamer in which four molecules are arranged tetrahedrally about a central molecule. They found that the water trimer in which the central water molecule was bonded through both the oxygen and hydrogen to the two other water molecules, was stabilized relative to the linear dimer by 0·5 kcal/mole per hydrogen bond. Their CNDO/2 calculation predicted that the pentamer had a stabilization energy of 5·81 kcal/mole per hydrogen bond, which is less than that obtained for the linear dimer. This result disagrees with experimental evidence which suggests that appreciable quantities of the pentamer (which is found in ice) should exist in liquid water. This disagreement may arise from the poor parameter set used, the approximations inherent in the method, and the fact that the geometry of the monomer was not varied on dimerization. With reference to this last point, Kollman and Allen[35] in their CNDO/2 study found that the calculated stabilization energy was rather sensitive to the geometry taken for the water monomer; this was in contrast to the *ab initio* results. Furthermore, experimental evidence[40] indicates that on dimerization the geometry of the monomer is distorted, the OH bond of the proton donor being lengthened.

Finally one can see from table 4.5 that the NDDO values are extremely poor. We believe this to be a reflection of a poor parameter set rather than any deficiencies in theory.

It is perhaps relevant to add that some stimulation for the examination of the energies of water clusters has arisen from the suggestion that a metastable 'polywater' had been prepared.[42] This appears now to lack experimental support, but CNDO calculations on some possible structures suggested cyclic hexamers as a possibility (stabilization 33·5 kcal/mole).[43]

(b) *Formic acid*

Considerable experimental evidence exists for the dimerization of carboxylic acids. Formic acid, the simplest member of this series, has been studied by several workers[34,44–46] using the CNDO/2 method with the parameter set of Pople and Segal.[38] Several possible conformations of the formic acid dimer and trimer were studied, some of the more obvious structures, with stabilization energies per H bond, are shown in fig. 4.8.

Calculations on the *cis* and *trans* conformations of the monomer were in agreement with the experimental[47] observation in predicting the *cis* conformation to be the more stable.[44,46] The Extended-Hückel method on the other hand predicted the *trans* form to be the preferred conformation.[44] The above calculations are also in accord in predicting the cyclic conforma-

Energies per H bond

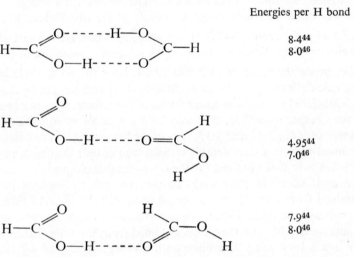

8.4[44]
8.0[46]

4.95[44]
7.0[46]

7.9[44]
8.0[46]

FIGURE 4.8   Structures and energies (kcal/mole) of the dimer of formic acid

tion to be the most stable structure of the formic acid dimer, although the energy difference between this and the open dimer is very small. These results are in agreement with the experimental evidence which indicates that formic acid exists in the gas phase as the cyclic dimer,[48] with a stabilization energy per hydrogen bond of about 7·0 kcal/mole.[40] Later experimental results indicated that the possibility of open structures for the dimer could not be ruled out.[49] Variation of the O–O distance produced minimum energies at 2·45 Å,[44] 2·40 Å[46] and 2·42 Å,[34] which are in rather poor agreement with the experimental distance of 2·73 Å.[48] The trimer of formic acid which is predicted by the CNDO/2 method to have a stabilization of 7·03 kcal/mole,[44] has been isolated and subjected to an X-ray crystallography study.[50] Acetic acid has also been studied with the CNDO/2 method[34] and the results show a stabilization energy per hydrogen bond of 11·8 kcal/mole (observed 7·5 kcal/mole[40]) and an O–O distance of 2·42 Å (observed 2·73 Å[48]) for the cyclic dimer.

4

(c) *Hydrogen Fluoride*

Hydrogen fluoride which has received considerable experimental attention has been studied with the all-valence-electron methods. CNDO/2[34,35], NDDO[35] and *ab initio*[35] calculations have been reported on various structures of some polymeric units of hydrogen fluoride.

Hoyland and Kerr[34] found increasing stabilization for the HF polymer as the number of monomer units increased. The stabilization energy of the dimer was calculated to be 6·92 kcal/mole at an equilibrium F–F distance of 2·43 Å (observed 2·45 Å[51]) and an equilibrium HFH angle of 145°. The energy of the dimer was found to be relatively insensitive to the HFH angle, hence the value of 145° was adopted for the bond angle in subsequent calculations on the polymer. Stabilization energies per hydrogen bond, calculated for the non-linear forms of the trimer, tetramer and hexamer were respectively 7·78, 8·27 and 8·82 kcal/mole at equilibrium F–F distances of 2·40, 2·39 and 2·37 Å. A cyclic hexamer was also calculated and found to have a considerable stabilization of 0·61 kcal/mole per hydrogen bond over that obtained for the open-chain analogue.

Kollman and Allen[35] in their study on the HF dimer compared the results obtained from CNDO/2, NDDO and *ab initio* calculations. All three methods predicted the linear form to be approximately twice as stable as the cyclic analogue, although the values obtained from the NDDO calculations were again very poor. The dimerization energy calculated by the CNDO/2 method was found to be rather sensitive to the geometry taken for the monomer. For example, experimental and calculated equilibrium geometries gave values which differed by 2·4 kcal/mole. The *ab initio* results on the other hand were relatively insensitive to the geometry taken for the monomer.

In a more extensive study of the hydrogen bonding in HF, by the CNDO/2 method, Kollman and Allen[35] considered a number of polymeric units. Their results, shown in table 4.6, were obtained with an HF bond length of 1·00 Å and an F–F distance of 2·45 Å; this last value was the calculated separation in the dimer.

Several interesting points emerge from a comparison of these results. Firstly, the stabilization energy gained per HF fragment reaches a maximum for the linear polymers at $(HF)_5$. Secondly, the cyclic polymer $(HF)_6$ is predicted to be the most stable polymer overall. Thirdly, the nonplanar cyclic structures are calculated to be less stable by approximately 1 kcal/mole, than the corresponding planar structure. These results are in agreement with an electron diffraction study[52] which found that hydrogen fluoride exists in the gas phase as a cyclic hexamer with an F–F distance of

TABLE 4.6 Stabilization energies per HF unit obtained from CNDO/2 calculations on some polymers of hydrogen fluoride (energies in kcal/mole)

	Linear Energy	Cyclic Structure	Energy
$(HF)_2$	4·505		—
$(HF)_3$	6·819		—
$(HF)_4$	8·117	square	10·45
$(HF)_5$	8·946	regular pentagon	10·70
$(HF)_6$	8·498[a]	regular hexagon	10·92
$(HF)_6$	—	cyclohexane boat structure	9·73
$(HF)_6$	—	cyclohexane chair structure	9·62
$(HF)_7$	8·553		—
$(HF)_8$	8·555	regular octagon	10·50
$(HF)_8$		cyclooctane structure	9·70

[a] a zigzag hexamer with $\widehat{FFF} = 160°$ was found to be less stable than the linear hexamer.

2·45 Å and $\widehat{FFF}$ angles of 104°. A geometry search was not carried out on any of the above structures, hence we cannot compare the calculated bond distances with those obtained experimentally. Kollman and Allen[35] also considered the effect on the molecular orbitals of the polymer, of an approaching HF molecule: there are two situations to be considered, depending on whether a fluorine or a hydrogen is the approaching atom,

$$HF \rightarrow HF \cdots HF$$

$$HF \cdots HF \leftarrow HF.$$

In the first case they found that the molecular orbitals of the polymer were raised in energy whilst in the second case the molecular orbitals were lowered in energy. In both of these cases a limiting value was reached when the addition of further HF fragments made little difference to the energies of the molecular orbitals of the polymer.

The SCF–MO calculations have therefore been reasonably successful in treating hydrogen-bonded systems. Several other systems have been examined, notably the dimers of formamide,[53] methanol,[34,44] ammonia,[34] the ammonium ion and the methylammonium ion–water system,[34] the formaldehyde–water system,[46] and the bifluoride and maleate ions.[54] Calculations were, however, unsuccessful for the biochemically important dimer of hydrogen cyanide.[34] Nevertheless, the results obtained so far on hydrogen-bonded systems must give one reasonable cause for optimism, particularly because all the CNDO/2 methods have used the parameter set of Pople and Segal[38] which, as we saw in chapter 3 does not represent the optimum set. Future studies in which a better parameter set is used would be expected to lead to considerably better results; this should be particularly true for the NDDO method. Although the SCF methods have failed to answer some of the important questions concerning hydrogen bonding, they have nevertheless provided chemists with some useful information.

4.4   Carbonium ions

It is well known that bimolecular nucleophilic substitution reactions ($S_N 2$) occur with inversion of configuration, and that a good model for the transition state is a trigonal bipyramidal structure.[55] Bimolecular electrophilic substitution reactions ($S_E 2$), on the other hand, were for a

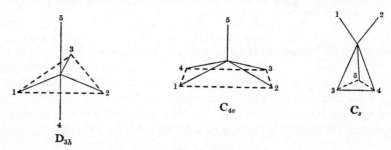

FIGURE 4.9   The most likely structures for CH_5^+

long time thought to occur exclusively with retention of configuration.[56] The problem of configuration in $S_E 2$ reactions has been studied with the SCF methods for a few examples. The most extensive study has been of the protonation of methane. This system has the additional interest that alkanes will protonate under extremely acidic conditions.[57] Methane, for example, will liberate hydrogen with the eventual formation of a polymeric hydrocarbon ion.[58,59] The experimental evidence suggested that a possible intermediate in the reaction was the CH_5^+ ion. Previously, this ion had been observed in the gas phase by mass spectrometry,[60] and its existence in the solid phase had been suggested by photolysis and radiolysis of methane.[61]

A considerable number of theoretical papers have been published on the possible structure of the CH_5^+ ion. In the main part these studies have dealt with the stability of the three most likely structures, which are shown in figure 4.9.

The stability of each structure has been determined with the CNDO/2 method,[62] using the parameter set of Wiberg[63] and by the PNDO method.[64] *Ab initio* calculations are also available.[65,66] Optimum geometries for each structure were determined, and in one calculation an extensive area of the potential-energy surface was examined.[66] Some results are shown in table 4.7.

TABLE 4.7 Calculated energies (kcal/mole) and geometries of CH_5^+

	D_{3h}			C_{4v}		C_s		
	CNDO/2	PNDO	*Ab initio*	CNDO/2	*Ab initio*	CNDO/2	PNDO	*Ab initio*
C—H_5(Å)	1·16	1·13	1·13	1·16	1·13	1·14	1·12	1·17
C—H_1(Å)	1·14	1·12	1·12	1·13	1·07	1·21	1·14	1·10
$\widehat{H_1CH_2}$(deg)				80	82	50	75	58
$\widehat{H_4CH_5}$(deg)						110	103	108
E_{Rel} kcal/ mole	9·7	−5·3	3·5	4·7	0·3	0·0	0·0	0·0

There appears to be no doubt that the potential-energy surface is very flat around the equilibrium configuration, as suggested by the small difference between the energies of the three structures. In addition, for the C_s structure there is essentially free rotation of H_1 and H_2 relative to the lower triangle of atoms, and furthermore it is very easy to extend the $CH_1(H_2)$ bonds. Other *ab initio* calculations on the C_s configuration give a substantially longer CH_1 bond (1·6 Å)[66] than than quoted in table 4.7.

The CNDO/2 and *ab initio* results are in agreement in predicting the following order of decreasing stability $C_s > C_{4v} > D_{3h}$. The PNDO results predict the D_{3h} configuration to be more stable than the C_s by 5·3 kcal/mole.

A PNDO study using the parameter set of Dewar and Klopman[67] has also been reported[59] on the CH_5^+ ion. No attempt was made to minimize the molecular energy with respect to geometry, since as we saw in chapter 3, this particular form of the PNDO method fails to give a correct estimate of the equilibrium geometry. Consequently, the calculations were carried out using standard bond lengths. Three configurations of the CH_5^+ ion were studied, two of which, the D_{3h} and C_s configurations were considered in the above calculations. The third configuration studied (A), was a structure in which the proton approached along the direction of one of the CH bonds. In agreement with the CNDO and *ab initio* results, this PNDO study predicted the C_s configuration to be the most stable, by a factor of 9 kcal/mole over that of the D_{3h} species and 34 kcal/mole over configuration

A. The calculated proton affinity of 7·76 eV is in poor agreement with the observed value of 4·95–5·58 eV.[68]

According to the overall conclusion of the above methods, electrophilic substitution in methane take place through the C_s intermediate with retention of configuration. However, two words of caution are in order here. Firstly, one cannot assume that all S_E2 reactions involving methane will take place with retention of configuration, since the reaction pathway will probably depend upon the nature of the attacking electrophilic

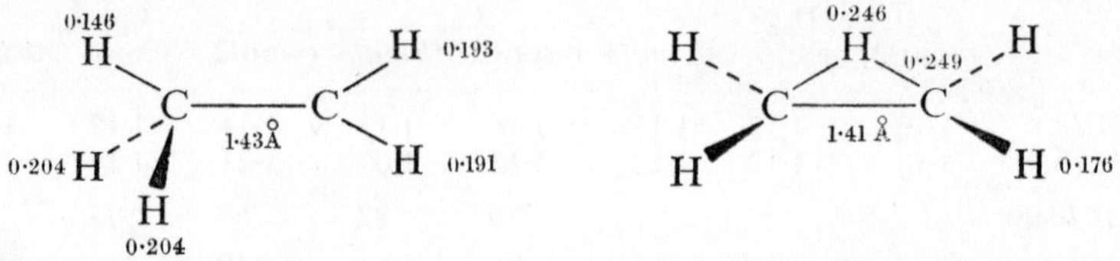

FIGURE 4.10   Possible structures for $C_2H_5^+$

reagent. Secondly, the above calculations refer to the gas phase and therefore we have no justification for extrapolating to solution where solvation energy, particularly for such a charged species, could be the dominant term.

A PNDO study into the nucleophilic substitution (S_N2) of methane by a hydride ion, has also been reported.[64] The results showed that the $\mathbf{D}_{3h}$ conformation was favoured over the $\mathbf{C}_s$ species by 14·9 kcal/mole.

Another simple carbonium system studied is the ion $C_2H_5^+$. An NDDO study has been reported,[69] the object of which was to examine the relative stability of two possible configurations corresponding to the ethyl cation I and protonated ethylene II shown in figure 4.10.

The energy of each configuration was minimized with respect to the C—C bond length; the C—H bond lengths were not varied, but given the values 1·093 Å and 1·084 Å for sp^3 and sp^2 carbon atoms respectively. Values obtained for the C—C bond length and charge densities are shown in figure 4.10.

Structure II was estimated to be more stable than I by 33·2 kcal/mole. A breakdown of the energy components for structure II revealed that although it had a much larger nuclear repulsion energy (300 kcal/mole), this was more than compensated by the gain in electronic energy. Inspection of the charge densities showed that there was a more even charge distribution in structure II than in I.

In recent years experimental evidence has accumulated in a number of reactions of the possible existence of a protonated cyclopropane ring

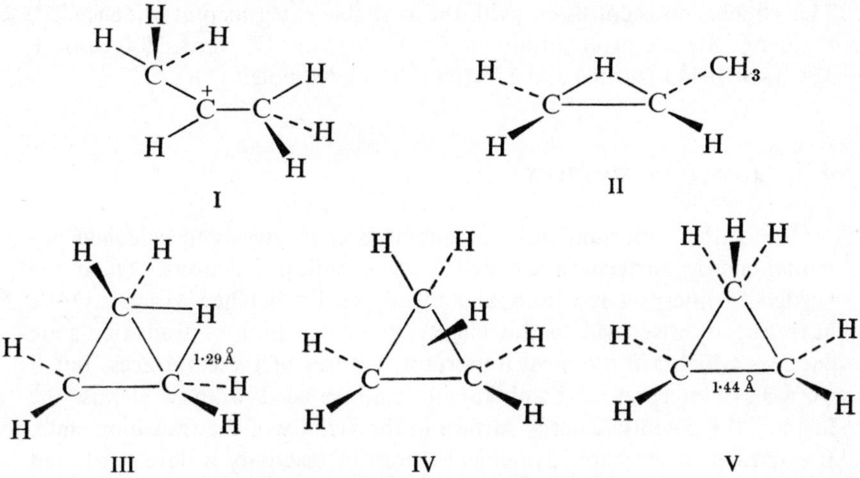

FIGURE 4.11   Possible structures for $C_3H_7^+$

structure.[70] Its existence in the gas phase had been proposed earlier, from mass spectrometric studies on the hydrocarbons.[71]

Theoretical studies into the structure of the $C_3H_7^+$ ion have been reported with the CNDO/2,[72] INDO,[72] NDDO[69] and *ab initio* methods.[73] Several possible structures were investigated and some of these are shown in figure 4.11.

A full geometry search was carried out with the CNDO/2 and INDO calculations.[72] In the NDDO calculations an assumed geometry was taken for each configuration. The *ab initio* calculations were performed on structures III and IV with the experimental geometry of cyclopropane, and the proton distance varied until the energy was minimized. Table 4.8 shows the energies obtained, relative to the edge-protonated structure of cyclopropane III.

There is unanimous agreement between the methods in predicting the edge-protonated structure of cyclopropane III to be the most stable.

TABLE 4.8   Energies (kcal/mole) of $C_3H_7^+$ structures relative to the edge protonated structure (III)

	CNDO/2[72]	INDO[72]	NDDO[69]	*Ab initio*[73]
I	—	—	81·0	—
II	—	—	63·7	—
IV	76	120	136·5	125·2
V	10	22	20·2	—

This conclusion is consistent with the available experimental evidence.[70,74] Estimates for the protonation energy of structure III were 235 (CNDO/2), 321 (INDO), 443 (NDDO) and 155 (*ab initio*) kcal/mole.

4.5 Reaction surfaces

One of the important tasks of quantum chemistry is to calculate potential-energy surfaces associated with chemical reactions. These are needed to interpret reactions and ideally to predict new reactions. We have already discussed in this chapter the calculation of transition-state energies which are the most important features of these surfaces, but if we wish to interpret $\Delta H^{\ddagger}$ and $\Delta S^{\ddagger}$ in detail we need to know at least the form of the potential-energy surface in the vicinity of the transition state. If eventually a workable dynamical theory of reactivity is developed then even more of the potential surface will be required.

The problem of calculating potential-energy surfaces is largely one of computing time. Suppose it takes a time T to calculate the energy for a given molecular configuration within a specified accuracy. If the molecule has N atoms then the potential-energy surface has $3N–6$ dimensions, X_i. If the energy is developed as a Taylor series about a point

$$E(X_i + \delta X_i) = E(X_i) + \sum_i \left(\frac{\partial E}{\partial X_i}\right) \delta X_i + \tfrac{1}{2} \sum_i \sum_j \left(\frac{\partial^2 E}{\partial X_i\, \partial X_j}\right) \delta X_i\, \delta X_j +,$$

$$(4.11)$$

then we can see that we need a minimum of $\tfrac{1}{2}(3N - 5)(3N - 4)$ points to construct the surface up to the harmonic terms. For the CH_5^+ ion for example this comes to 91 points, or a minimum time of 91 T. Unfortunately a harmonic surface will not be accurate over a very wide area. In practice one looks for some aspects of symmetry that might simplify the problem, which is why only configurations of relatively high symmetry have been examined for the CH_5^+ ion.

A second problem that one must face is that the type of calculation which gives reasonably good relative energies for molecules at their equilibrium configurations may give poor energies at other points on the reaction surface. The problem encountered here is one of correlation energy, which it is not our intention to discuss in detail. We do, however, emphasize the well-known fact that molecular orbital theory in its usual closed-shell form, with electrons occupying orbitals in pairs, gives poor dissociation energies unless some correction is made for the correlation energy of two electrons in the same orbital.

To have any confidence that a particular method of calculation would be accurate over a potential-energy surface one would require that it at least reproduces the energies, geometries and force constants of *stable* molecules with reasonable accuracy. At the present time the empirical SCF theory which most nearly meets this criterion is the MINDO/2 method of Dewar and coworkers.[75,76] We will describe some of its applications to reaction surfaces.

(a) *Rotation about* C$=$C *double bonds*

One of the earliest studies made[75] was an investigation into the barrier of rotation about C$=$C double bonds. Five molecules were considered and the

TABLE 4.9 MINDO/2 results for some systems containing C$=$C bonds (energies in kcal/mole)[75]

	Heat of formation		C—C bond lengths Å		Barrier to rotation		Torsional frequency (cm^{-1})	
	Calc.	Obs.	Calc.	Obs.	Calc.	Obs.	Calc.	Obs.
Ethylene	16·4	12·5	1·337	1·337	54·1	65·0	1094	1027
Allene	42·7	45·9	1·309	1·308	35·2		929	865
Butatriene	72·4		1·311, 1·288	1·318, 1·283	32·3	30	841	736
Pentatetraene	99·0		1·310, 1·288		26·1		747	

total energy was minimized with respect to geometry at each angle θ of rotation. Table 4.9 shows some of the results obtained, together with the corresponding experimental values.

Calculated rotation barriers and torsional frequencies are in fairly good agreement with experiment. A plot of the calculated energy against θ, showed that the form of the curve did not follow the expression

$$E(\theta) = E_0 + V_0 \sin \theta, \qquad (4.12)$$

which has usually been assumed for this process. Dewar and Haselbach[75] pointed out that this is understandable because the torsional frequency of ethylene which is calculated from a sine function (4.12), for a barrier height of 54·1 kcal/mole, is 1235 cm^{-1}, as compared with the experimental value of 1027 cm^{-1}.[77]

An *ab initio* calculation[78] on the barrier to rotation in ethylene gave a barrier height of 82·1 kcal/mole.

TABLE 4.10   MINDO/2 results for cyclooctatetraene (COT) and some of its ions

	CĈC	Alternating bonds Bond lengths (Å)		Heat of formation kcal/mole	Equal bonds Bond lengths (Å)	Heat of formation kcal/mole
COT⁻⁻, planar					1·418	128·8
COT⁻, planar		1·383	1·442	44·3	1·413	47·9
nonplanar	132°	1·381	1·442	47·1	1·413	47·1
COT, planar		1·355	1·475	73·7	1·410	87·6
nonplanar	125°	1·355	1·473	56·7	1·407	71·3
COT⁺, planar		1·381	1·440	248·7	1·410	252·2
nonplanar	130°	1·379	1·442	245·2	1·409	249·0
COT⁺⁺, planar					1·414	538·0

(b) *Ring inversion and bond shift in cyclooctatetraene*

Ring inversion and the associated double-bond shift in cyclooctatetraene has been studied[79] with the MINDO/2 method. The energy of cyclooctatetraene and some of its ions, was minimized as a function of the C—C bond lengths for a structure with alternating single and double bonds, and for structures in which all bond lengths were made equal. The results are shown in table 4.10.

The results indicate cyclooctatetraene to be most stable as a non-planar (boat) structure with alternating bonds, in agreement with experiment.

Ring inversion in cyclooctatetraene has been studied experimentally by Anet and coworkers[80] and by Roberts and coworkers,[81] both groups concluding that inversion takes place without single–double bond shift. The above results are consistent with this observation, because the planar form of cyclooctatetraene, which is the transition state in ring inversion, is predicted to retain the alternating bond structure. The calculated activation energy of 17 kcal/mole for inversion is in satisfactory agreement with the experimental value of 12–15 kcal/mole.[80,81]

Anet and coworkers[80] further suggested that bond shift would take place through the symmetrical planar form of cyclooctatetraene with an estimated difference of 2·4 kcal/mole between free energy of activation for ring inversion and for bond shift. The above results do not support this conclusion, since they predict an appreciable energy difference (13·9 kcal/mole) between the symmetrical and alternating forms of the planar structure.

Further calculations on cyclooctatetraene were made in which the energy of the non-planar form of the symmetrical and alternating structures were calculated as a function of the CĈC bond angle. This showed that bond angle had little effect on the molecular energy. It was concluded from this that bond shift would take place more readily through a non-planar than a planar structure, because a transition state involving a planar structure would be less favoured by the energy required to achieve planarity.

The monoanions of cyclooctatetraene were predicted to be non-planar, and the dianion and dication to be planar. It is known[82] that the monoanion of cyclooctatetraene is unstable to disproportionation according to the following reaction

$$2COT^- \rightarrow COT + COT^{--}.$$

However, the results given in table 4.10 show this reaction to be endothermic by 91·3 kcal/mole, so that if the MINDO/2 results are to be believed then the reaction must be driven by the greater solvation energy of the cyclooctatetraene dianion.

(c) Insertion reactions

MINDO/2 calculations have been carried out on the insertion of carbon atoms into double bonds to form allenes.[83] The reaction surfaces for insertion into ethylene and *trans*-2-butene have been studied and the energy along the reaction coordinate for the ethylene case is shown in figure 4.12.

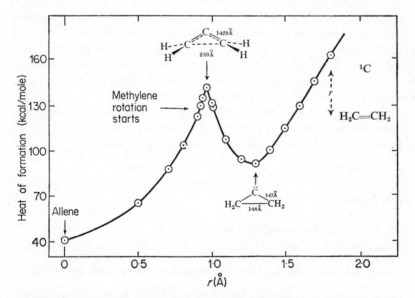

FIGURE 4.12 Reaction path calculated with MINDO/2 for the insertion of carbon atoms into ethylene[83]

The reaction coordinate is the distance from the incoming carbon atom to the centre of the C=C bond. The energy has been minimized with respect to all other coordinates.

It can be seen that the first process is the formation of a stable intermediate carbene. This then requires a high activation energy of 50 kcal/mole for its conversion to allene. The conversion of the intermediate to the allene requires that the methylene groups in the intermediate be rotated by 90°, since they are in mutually perpendicular planes in allene. The calculation shows that this rotation occurs after the transition state. These conclusions appear to be at variance with the experimental evidence,[84] which suggests that ^1S carbon atoms insert into double bonds at $-190°C$.

A possible explanation for the facile conversion of the carbene to the allene is that the conversion occurs before the excess vibrational energy of the carbene is dissipated. However, other reactions which would be expected to give cyclopropylidenes as intermediates also give allenes under mild conditions.[85]

Summary

Although the MINDO/2 method represents the most satisfactory empirical SCF–MO method to date of calculating reaction surfaces it by no means gives in all cases energies of the desired accuracy. For example, small rings are predicted to be too stable by approximately 10 kcal/mole and the C—H bond lengths are consistently too long by 0·1 Å.† Nevertheless the MINDO/2 method gives us some indication of what we might expect in the future. Considerably greater accuracy might be expected from a correctly parameterized NDDO method. Chemists may then be in a position of calculating reaction surfaces with a fair degree of confidence, and consequently be able to choose between several possible mechanistic pathways.

References

1. E. Hückel, *Z. Physik*, **70**, 204 (1931).
2. H. J. Dauben Jr., and D. J. Bertelli, *J. Amer. Chem. Soc.*, **83**, 4659 (1961).
3. R. S. Mulliken and R. G. Parr, *J. Chem. Phys.*, **19**, 1271 (1951).
4. C. A. Coulson and S. L. Altmann, *Trans. Faraday Soc.*, **48**, 293 (1952).

† In a private communication from Professor Dewar he informs us that this last difficulty has now been overcome.

5. M. J. S. Dewar, *Tetrahedron* (*suppl.*), **8**, 75 (1966).
6. M. J. S. Dewar and G. J. Gleicher, *J. Amer. Chem. Soc.*, **87**, 692 (1965).
7. M. J. S. Dewar and C. de Llano, *J. Amer. Chem. Soc.*, **91**, 789 (1969).
8. M. J. S. Dewar, A. J. Harget and N. Trinajstic, *J. Amer. Chem. Soc.*, **91**, 6321 (1969).
9. B. Bak, L. Hansen and J. Rastrup-Andersen, *Discuss. Faraday Soc.*, **19**, 30 (1955).
10. E. Clar, *Tetrahedron*, **5**, 98 (1959); **6**, 355.
11. N. C. Baird, *Can. Journ. of Chem.*, **47**, 3535 (1969).
12. D. H. Lo and M. A. Whitehead, *Can. Journ. of Chem.*, **46**, 2027, 2041 (1968).
13. G. W. Wheland, *J. Amer. Chem. Soc.*, **64**, 900 (1942).
14. A. Streitwieser Jr., *Molecular Orbital Theory for Organic Chemists*, Wiley, New York, London, 1961.
15. O. Chalvet, R. Daudel and J. J. Kaufman, *J. Phys. Chem.*, **68**, 490 (1964).
16. M. J. S. Dewar and C. C. Thompson Jr., *J. Amer. Chem. Soc.*, **87**, 4414 (1965).
17. L. P. Hammett, *Physical Organic Chemistry*, McGraw-Hill, New York, 1940.
18. M. Schwartz, *Chem. Soc. Special Publication*, **16**, 25 (1962).
19. A. Streitwieser Jr. and W. C. Langworthy, *J. Amer. Chem. Soc.*, **85**, 1757, 1761 (1963).
20. M. J. S. Dewar and R. J. Sampson, *J. Chem. Soc.*, **1956**, 2789; **1957**, 2946.
21. L. Verbit and E. Berliner, *J. Amer. Chem. Soc.*, **86**, 3307 (1964).
22. L. Altschuler and E. Berliner, *J. Amer. Chem. Soc.*, **88**, 5837 (1966).
23. G. J. Gleicher, *J. Amer. Chem. Soc.*, **90**, 3397 (1968).
24. A. Kende, *Advan. Chem. Physics*, **8**, 133 (1965).
25. M. J. S. Dewar and T. Morita, *J. Amer. Chem. Soc.*, **91**, 802 (1969).
26. R. L. Flurry Jr., and P. G. Lykos, *J. Amer. Chem. Soc.*, **85**, 1033 (1963).
27. J. D. Vaughan, D. C. Fullerton and C. Chang, *Internat. J. Quantum Chem.*, **2**, 205 (1968).
28. T. P. Lewis, *Tetrahedron*, **25**, 4117 (1969).
29. R. G. Jesaitis and A. Streitwieser Jr., *Theoret. Chim. Acta*, **17**, 165 (1970).
30. N. C. Baird, *Can. Journ. of Chem.*, **47**, 2306 (1969).
31. A. Streitwieser Jr., P. C. Mowery, R. G. Jesaitis and A. Lewis, *J. Amer. Chem. Soc.*, **92**, 6529 (1970).
32. J. N. Murrell, *Chem. in Britain*, **5**, 107 (1969).
33. S. Bratoz, *Adv. Quant. Chem.*, **3**, 209 (1967).
34. J. R. Hoyland and L. B. Kier, *Theoret. Chim. Acta*, **15**, 1 (1969).
35. P. A. Kollman and L. C. Allen, *J. Amer. Chem. Soc.*, **92**, 753 (1970).
36. K. Morokuma and L. Pedersen, *J. Chem. Phys.*, **48**, 3275 (1968).
37. A. S. N. Murthy and C. N. R. Rao, *Chem. Phys. Letters*, **2**, 123 (1968).
38. J. A. Pople and G. A. Segal, *J. Chem. Phys.*, **44**, 3289 (1966).
39. J. E. Harries, W. J. Burroughs, H. A. Gebbie, *J. Quant. Spectrosc. Radiat. Transfer*, **9**, 799 (1969).
40. G. C. Pimental and A. L. McClellan, *The Hydrogen Bond*, Freeman, San Francisco and London, 1960.
41. R. J. Jakobsen and J. W. Brasch, *Spectrochim. Acta*, **21**, 1753 (1965).
42. B. V. Deryagin and N. V. Churaev, *Priroda*, **1968**, 16 (translated in Joint Publication Research Service, **1968**, 45).
43. L. C. Allen and P. A. Kollman, *Science*, **167**, 1443 (1970).

44. A. S. N. Murthy, R. E. Davis and C. N. R. Rao, *Theoret. Chim. Acta*, **13**, 81 (1969).
45. P. Schuster and Th. Funck, *Chem. Phys. Letters*, **2**, 587 (1968).
46. P. Schuster, *Internat. J. Quantum. Chem.*, **3**, 851 (1969).
47. T. Miyazawa and K. S. Pitzer, *J. Chem. Phys.*, **30**, 1076 (1959).
48. J. Karle and L. O. Brockway, *J. Amer. Chem. Soc.*, **66**, 574 (1944).
49. P. Waldstein and L. A. Blatz, *J. Phys. Chem.*, **71**, 2271 (1967).
50. F. Holtzberg, B. Post and I. Fankuchen, *Acta Cryst.*, **6**, 127 (1953).
51. S. H. Bauer, J. Y. Beach and J. H. Simons, *J. Amer. Chem. Soc.*, **61**, 19 (1939).
52. J. Janzen and L. S. Bartell, *J. Chem. Phys.*, **50**, 3611 (1969).
53. A. Pullman and H. Berthod, *Theoret. Chim. Acta*, **10**, 461 (1968).
54. A. S. N. Murthy, S. N. Bhat and C. N. R. Rao, *J. Chem. Soc.* (A), **1970**, 1251.
55. C. K. Ingold, *Structure and Mechanism in Organic Chemistry*, Cornell University Press, Ithaca, N.Y. 1953.
56. F. R. Jensen and B. Rickborn, *Electrophilic Substitution of Organo-mercurials* McGraw-Hill, New York, 1968.
57. G. A. Olah and J. Lukas, *J. Amer. Chem. Soc.*, **89**, 2227, 4743 (1967).
58. G. A. Olah and R. H. Schlosberg, *J. Amer. Chem. Soc.*, **90**, 2726 (1968).
59. G. A. Olah, G. Klopman and R. H. Schlosberg, *J. Amer. Chem. Soc.*, **91**, 3261 (1969).
60. F. H. Field and M. S. B. Munson, *J. Amer. Chem. Soc.*, **87**, 3289 (1965) and refs. therein.
61. P. Ausloos, R. E. Rebbert and S. G. Lias, *J. Chem. Phys.*, **42**, 540 (1965).
62. A. Gamba, G. Morosi and M. Simonetta, *Chem. Phys. Letters*, **3**, 20 (1969).
63. K. B. Wiberg, *J. Amer. Chem. Soc.*, **90**, 59 (1968).
64. N. L. Allinger, J. C. Tai and F. T. Wu, *J. Amer. Chem. Soc.*, **92**, 579 (1970).
65. W. Th. A. M. van der Lugt and P. Ros, *Chem. Phys. Letters*, **4**, 389 (1969).
66. M. F. Guest, J. N. Murrell and J. B. Pedley, *Mol. Phys.*, **20**, 81 (1971).
67. M. J. S. Dewar and G. Klopman, *J. Amer. Chem. Soc.*, **89**, 3089 (1967).
68. V. L. Tal'rose and E. L. Frankevitch, *J. Amer. Chem. Soc.*, **80**, 2344 (1958).
69. R. Sustmann, J. E. Williams, M. J. S. Dewar, L. C. Allen and P. von R. Schleyer, *J. Amer. Chem. Soc.*, **91**, 5350 (1969).
70. C. C. Lee and J. E. Kruger, *Tetrahedron*, **23**, 2539 (1967) and refs. therein; C. C. Lee and L. Gruber, *J. Amer. Chem. Soc.*, **90**, 3775 (1968).
71. P. N. Rylander and S. Meyerson, *J. Amer. Chem. Soc.*, **78**, 5799 (1956).
72. H. Fischer, H. Kollmar and H. O. Smith, *Tet. Letters*, **1968**, 5821.
73. J. D. Petke and J. L. Whitten, *J. Amer. Chem. Soc.*, **90**, 3338 (1968).
74. C. J. Collins, *Chem. Rev.*, **69**, 543 (1969).
75. M. J. S. Dewar and E. Haselbach, *J. Amer. Chem. Soc.*, **92**, 590 (1970).
76. N. Bodor, M. J. S. Dewar, A. J. Harget and E. Haselbach, *J. Amer. Chem. Soc.*, **92**, 3854 (1970).
77. R. L. Arnett and B. L. Crawford Jr., *J. Chem. Phys.*, **18**, 118 (1950).
78. R. J. Buenker, *J. Chem. Phys.*, **48**, 1368 (1968).
79. M. J. S. Dewar, A. J. Harget and E. Haselbach, *J. Amer. Chem. Soc.*, **91**, 7521 (1969).
80. F. A. L. Anet, *J. Amer. Chem. Soc.*, **84**, 671 (1962). F. A. L. Anet, A. J. R. Bourn and Y. S. Lin, *J. Amer. Chem. Soc.*, **86**, 3576 (1964).

81. D. E. Gwynn, G. M. Whitesides and J. D. Roberts, *J. Amer. Chem. Soc.*, **87,** 2862 (1965).
82. T. J. Katz, *J. Amer. Chem. Soc.*, **82,** 3784, 3785 (1960).
83. M. J. S. Dewar, E. Haselbach and M. Shanshal, *J. Amer. Chem. Soc.*, **92,** 3505 (1970).
84. P. S. Skell and R. R. Engel, *J. Amer. Chem. Soc.*, **89,** 2912 (1967).
85. W. von E. Doering and P. M. LaFlamme, *Tetrahedron*, **2,** 75 (1958).

Chapter 5

Magnetic resonance spectroscopy

5.1 Introduction

Magnetic resonance spectroscopy has in the last twenty years become not only an important analytical tool of the chemist but also a source of data for testing theories of electronic structure. A necessary condition for the experiment is to have a molecule with a degenerate state whose degeneracy is removed in the presence of a magnetic field. Electromagnetic radiation of the appropriate frequency will then stimulate transitions between the components of the state.

As an example we can take a single spin-$\frac{1}{2}$ particle (for example a proton, or an electron) which has two spin states. In the presence of a uniform magnetic field H the energy levels are μH and $-\mu H$ where μ is the magnetic moment of the particle. In qualitative terms these correspond to states in which the magnetic moment of the particle is parallel and antiparallel to the magnetic field. A transition between the two states can be stimulated by the magnetic component of electromagnetic radiation, provided that the Planck–Einstein resonance condition is satisfied.

$$h\nu = \Delta E = 2\mu H. \tag{5.1}$$

The situation is illustrated in figure 5.1. The experiment can be carried out either by sweeping ν for a fixed H, or by sweeping H for a fixed ν; both types of spectrometers are widely available.

From electron spin resonance (ESR) or electron paramagnetic resonance (EPR) one determines the distribution of unpaired electrons in the molecule and the effective magnetic moment of the electrons arising from the combination of their spin and orbital angular momenta. There is also a coupling between the electrons and any nuclei in the molecule having magnetic moments. In nuclear magnetic resonance, NMR, one determines the field H actually experienced by the nucleus, which is different from the external field because of the shielding effect of the electrons. The spectra also depend on the mutual interaction or coupling of different nuclei as transmitted by the electrons.

Both techniques need a knowledge of molecular electronic structure for their complete understanding. However, as the theory involved is quite different for the two cases we will consider them separately. For a more

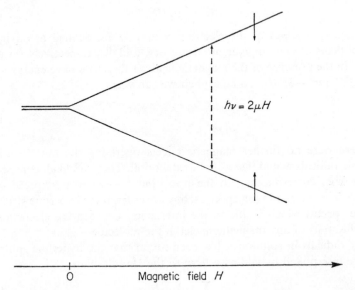

$$hv = 2\mu H$$

O   Magnetic field H

FIGURE 5.1   The magnetic resonance experiment for a single spin-$\frac{1}{2}$ particle

comprehensive treatment of the theory of magnetic resonance we recommend the book by Carrington and McLachlan.[1]

5.2   The ESR spectra of organic free radicals in solution

The calculations we shall describe have been made on organic radicals and are used to interpret the ESR spectra of those radicals in solution. The molecules contain no heavy atoms and hence spin–orbit coupling is unimportant. The magnetic moment of the electron is therefore composed only of the spin contribution which we write

$$\mu = -g\beta S, \qquad (5.2)$$

S being the spin angular momentum operator. β is the Bohr magneton equal to $eh/2mc$ and g is a constant called the electron g factor: $2 \cdot 0023$.

In a uniform magnetic field the interaction energy is given by the so-called Zeeman Hamiltonian

$$\mathscr{H} = -\mu \cdot \mathbf{H} = g\beta \mathbf{S} \cdot \mathbf{H}, \tag{5.3}$$

and for a magnetic field in the z direction

$$\mathscr{H} = g\beta H S_z. \tag{5.4}$$

The two spin states of the electron, α and β, are defined as having z-components of spin angular momenta of $\frac{1}{2}$ and $-\frac{1}{2}$ respectively (in units of $\hbar$). In the presence of the magnetic field we therefore have energy levels of $\frac{1}{2}g\beta H$ and $-\frac{1}{2}g\beta H$. To be explicit we can write

$$\begin{aligned}
\langle \alpha | \mathscr{H} | \alpha \rangle &= \tfrac{1}{2}g\beta H \\
\langle \beta | \mathscr{H} | \beta \rangle &= -\tfrac{1}{2}g\beta H.
\end{aligned} \tag{5.5}$$

If there were no further magnetic effects operating the ESR spectra of organic radicals would therefore be rather dull. They would all consist of a single line, corresponding to the transition $\beta \to \alpha$, at a frequency $v = g\beta H/h$. The interest in ESR spectroscopy arises from the hyperfine structure in the spectra which is due to the interaction between the electron spin and the spins of any magnetic nuclei in the molecule.

For radicals in solution it has been found that the hyperfine splittings can be interpreted by adding a term to the Hamiltonian

$$\sum_{N} a_N \mathbf{I}_N \cdot \mathbf{S} \tag{5.6}$$

where $\mathbf{I}_N$ is the angular momentum operator of nucleus N and $\mathbf{S}$ is the total electron spin angular momentum:

$$\mathbf{S} = \sum_{k} \mathbf{S}_k.$$

The hyperfine constants a_N may be determined from the spectra, as we shall see, and then interpreted on the basis of calculated electronic wave functions.

To show how a is determined from a spectrum we consider the simplest example of a radical with one spin-$\frac{1}{2}$ magnetic nucleus. We can represent the combined spin states of the electron and nucleus by the products $\alpha\alpha$, $\alpha\beta$, $\beta\alpha$ and $\beta\beta$, where the first symbol refers to the nucleus and the second to the electron.

Because the magnetic moments of nuclei are about one thousand times smaller than the magnetic moment of the electron, we can neglect the contribution to the energy from the nuclear Zeeman term, that is the term $-g_N\beta_N\mathbf{H}\cdot\mathbf{I}$, which is analogous to (5.3). The effective Hamiltonian is therefore from (5.4) and (5.6)

$$\mathscr{H} = g\beta H S_z + a\mathbf{I}\cdot\mathbf{S}. \tag{5.7}$$

The electron Zeeman energy has already been calculated, and to this we need to add the hyperfine energy. We note that

$$a\mathbf{I}\cdot\mathbf{S} = a(I_z S_z + I_x S_x + I_y S_y) \tag{5.8}$$

and from the fact that our spin states have well-defined z components, we have

$$\langle\alpha\alpha| \, aI_z S_z \, |\alpha\alpha\rangle = \langle\beta\beta| \, aI_z S_z \, |\beta\beta\rangle = \quad a/4$$
$$\langle\alpha\beta| \, aI_z S_z \, |\alpha\beta\rangle = \langle\beta\alpha| \, aI_z S_z \, |\beta\alpha\rangle = -a/4. \tag{5.9}$$

The terms $I_x S_x$ and $I_y S_y$ in (5.8) make no contribution to the diagonal elements of the Hamiltonian matrix because when operating on one spin state they convert it to the other. This is more exactly stated by writing

$$I_x S_x + I_y S_y = \tfrac{1}{2}[(I_x + iI_y)(S_x - iS_y) + (I_x - iI_y)(S_x + iS_y)]$$
$$\equiv \tfrac{1}{2}(\mathbf{I}^+\mathbf{S}^- + \mathbf{I}^-\mathbf{S}^+), \tag{5.10}$$

and using the following relationships for these so-called shift operators:

$$\mathbf{S}^+\,|\alpha\rangle = 0, \quad \mathbf{S}^-\,|\alpha\rangle = |\beta\rangle,$$
$$\mathbf{S}^+\,|\beta\rangle = |\alpha\rangle, \quad \mathbf{S}^-\,|\beta\rangle = 0. \tag{5.11}$$

There are corresponding expressions for the nuclear operators $\mathbf{I}^+$ and $\mathbf{I}^-$. We therefore see for example that

$$\langle\alpha\beta| \, a(I_x S_x + I_y S_y) \, |\alpha\beta\rangle = \langle\alpha\beta| \, a/2 \, |\beta\alpha\rangle = 0 \tag{5.12}$$

because α and β are orthogonal spin functions.

The only non-zero off-diagonal element of $a\mathbf{I}\cdot\mathbf{S}$ is

$$\langle\alpha\beta| \, a\mathbf{I}\cdot\mathbf{S} \, |\beta\alpha\rangle = a/2 \tag{5.13}$$

there being in this a contribution from $\mathbf{I}^+\mathbf{S}^-$.

The usual experimental conditions are such that the difference in Zeeman-energy between the state $\alpha\beta$ and $\beta\alpha$ is very much larger than the matrix element (5.13). It follows that this off-diagonal interaction only leads to small second-order energy changes which cannot usually be detected experimentally. We note therefore, in passing, that the spin-Hamiltonian for ESR may be taken in the simpler form

$$\mathscr{H} = g\beta H S_z + \sum_N a_N I_{zN} S_z. \tag{5.14}$$

The electron-spin transitions are allowed only between states with the same nuclear spin (that is a photon can only turn over the spin of one particle at a time). These are shown in figure 5.2. The spectrum therefore consists of two lines separated by a, whose mean position is $g\beta H$.

The hyperfine splitting pattern for many nuclear spins is easily derived from this type of analysis. The following is a summary of the position, and further details can be found in the book by Carrington and Mc-Lachlan.[1]

1. A spin-$\frac{1}{2}$ nucleus splits the free electron line into two components with separation a_1.

2. A second spin-$\frac{1}{2}$ nucleus further splits each of these lines into two with separation a_2.

3. When $a_2 = a_1$ it can be seen that the resulting pattern is of three lines separated by a_1 with relative intensities in the ratio $1:2:1$.

4. Spin-$\frac{1}{2}$ nuclei with the same a factors split the free electron line into

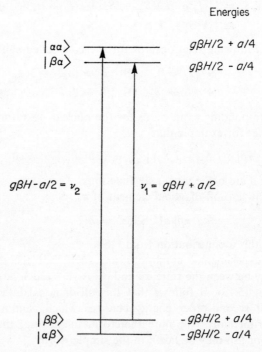

FIGURE 5.2 Allowed electron-spin transitions for a radical with one spin-$\frac{1}{2}$ nucleus

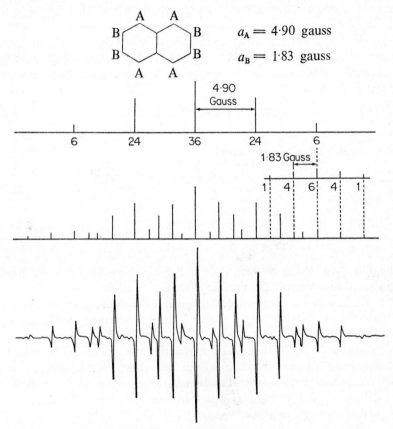

$a_A = 4\cdot90$ gauss

$a_B = 1\cdot83$ gauss

FIGURE 5.3 The E.S.R. spectrum of the naphthalene anion and its analysis.[1] The two splitting constants are $a_A = 4\cdot90$, $a_B = 1\cdot83$ gauss. These splittings may, if desired, be converted into frequency units (Mc/s) simply by multiplication by $2\!:\!80$. Note that this is *only* true for $g = 2$

$n + 1$ equally spaced lines of separation a, whose intensities are in the ratio of the binomial numbers. That is, the rth line has an intensity proportional to the coefficient of x^r in the expansion of $(1 + x)^n$.

5. A spin–I nucleus splits the free electron line into $(2I + 1)$ components of equal intensity, whose separation is a.

By a suitable combination of these rules one can elucidate any splitting pattern.

Figure 5.3 shows the high resolution ESR spectra of the naphthalene negative ion and its analysis. For reasons of sensitivity ESR spectrometers usually record the derivative of the absorption curve, as is shown in the figure.

We turn now to the interpretation of the a factors through the electronic wave functions. The interaction between electron and nuclear spins has two parts. Firstly there is the dipole–dipole interaction, which in vector notation is written

$$\frac{\mathbf{\mu}_e \cdot \mathbf{\mu}_N}{r_N^3} - \frac{3(\mathbf{\mu}_e \cdot \mathbf{r}_N)(\mathbf{\mu}_N \cdot \mathbf{r}_N)}{r_N^5}, \tag{5.15}$$

or in terms of the spin operators

$$-g\beta g_N\beta_N\left\{\frac{\mathbf{I}_N \cdot \mathbf{S}}{r_N^3} - \frac{3(\mathbf{I}_N \cdot \mathbf{r}_N)(\mathbf{S} \cdot \mathbf{r}_N)}{r_N^5}\right\}. \tag{5.16}$$

By analogy with (5.2) we are here writing

$$\mathbf{\mu}_N = g_N\beta_N\mathbf{I}_N, \tag{5.17}$$

the difference in sign between (5.2) and (5.17) being due to the difference in sign of the electron and nuclear charges.

For a rapidly tumbling molecule, which is the situation in solution, the dipole–dipole energy averages to zero for all spin states of the electron and nucleus.

Expression (5.16) is inadequate to describe the interaction between the electron and nucleus when the two are in contact, that is when $\mathbf{r}_N = 0$. The additional term in the Hamiltonian which is found to explain the observations, and which was first introduced by Fermi to account for hyperfine structure in atomic spectra is

$$\mathscr{H}' = \frac{8\pi}{3} g\beta g_N\beta_N \sum_k \delta(\mathbf{r}_{kN})\mathbf{I}_N \cdot \mathbf{S}_k. \tag{5.18}$$

It involves a scalar product of the electron and nuclear spin angular momenta operators but is only effective when the electron has a finite probability of being at the nucleus. This factor is introduced through the delta function $[\delta(\mathbf{r}_{kN})]$ which is defined in the operator sense of

$$\int \psi \, \delta(\mathbf{r}_{kN})\psi \, \mathrm{d}v = \psi^2(r_{kN} = 0) \tag{5.19}$$

being the probability density for the electron k at nucleus N. Thus if ψ is an atomic orbital, the operator $\mathscr{H}'$ only has non-zero expectation values for s orbitals, other orbitals (p, d etc.) having nodal planes passing through the nucleus.

In order to relate the hyperfine constant a to an expectation value of the Hamiltonian (5.18) we shall equate matrix elements of (5.6) and (5.18). We have seen earlier that only the diagonal elements of the matrix need be considered to interpret the spectra, and indeed the relationship we shall derive only holds for diagonal elements. We consider a radical in an electron-spin eigenstate with $S_z = \frac{1}{2}$. We represent this electronic state, which is composed of a space and spin part, by Ψ_e. The nuclear part we represent by Ψ_n. We wish to establish the identity

$$\frac{1}{2}a_N\langle\Psi_n|\,I_{zN}\,|\Psi_n\rangle = \frac{8\pi}{3}\,g\beta g_N\beta_N\langle\Psi_e\Psi_n|\sum_k \delta(\mathbf{r}_{kN})I_{zN}S_{zk}\,|\Psi_e\Psi_n\rangle. \quad (5.20)$$

The nuclear part can be factored out leaving

$$\frac{a_N}{2} = \frac{8\pi}{3}\,g\beta g_N\beta_N\langle\Psi_e|\sum_k \delta(\mathbf{r}_{kN})S_{zk}\,|\Psi_e\rangle. \quad (5.21)$$

From the definition of the delta function in (5.19) we note that

$$\int\Psi_e\left(\sum_k \delta(\mathbf{r}_{kN})\right)\Psi_e\,d\tau \quad (5.22)$$

is the total electron density at nucleus N. If in this total density we count α electrons as positive and β as negative, then we get a quantity which can be called the spin density at nucleus N. This can be written

$$\rho(r_N) = \int\Psi_e\left(\sum_k 2S_{zk}\,\delta(\mathbf{r}_{kN})\right)\Psi_e\,d\tau. \quad (5.23)$$

Thus from (5.21) and (5.23) we have

$$a_N = \frac{8\pi}{3}\,g\beta g_N\beta_N\rho(r_N), \quad (5.24)$$

Thus to calculate the hyperfine constant we need only to calculate the spin-density of the radical at the nucleus in question.

In the simplest type of wave function for a radical we assign electrons to molecular orbitals, in pairs, and have one molecular orbital with an unpaired electron. This is described by the Slater determinant (see appendix 2)

$$\Psi = |\psi_{1\alpha}\psi_{1\beta}\cdots\psi_{m\alpha}\psi_{m\beta}\psi_{n\alpha}|. \quad (5.25)$$

The electrons in orbitals $\psi_1-\psi_m$ make no contribution to the spin density because the α and β electron densities are equal. The spin density therefore is associated solely with orbital ψ_n, and we have

$$\rho(r_N) = \int\psi_n\,\delta(\mathbf{r}_N)\psi_n\,dv. \quad (5.26)$$

If we have an LCAO expansion for ψ_n

$$\psi_n = \sum_\mu c_{n\mu} \varphi_\mu, \qquad (5.27)$$

and there is just one s orbital associated with atom N, then we can approximate (5.26) by

$$\rho(r_N) = c^2{}_{nN} s^2{}_N(0) \qquad (5.28)$$

$s_N{}^2(0)$ being the s-orbital density at the nucleus. In this approximation we are neglecting all electron densities at nucleus N from orbitals of other atoms; an approximation consistent with a ZDO model. We shall later discuss a more general single-determinant wave function than (5.25).

The class of radicals whose ESR spectra have been most widely studied in solution are those obtained from planar aromatic hydrocarbons. These contain the radical anions and cations, like the anthracene anion and cation, and neutral radicals like benzyl. Many radicals of this type have also been studied with substituent groups.

For all of these aromatic radicals the first approximation to the wave function places the unpaired electron in a π orbital. Because such an orbital has a nodal plane passing through the conjugated atoms, and through the nuclei of any hydrogen atoms attached to the ring, the spin density at such nuclei would, in this approximation, be zero. However, hyperfine splittings are observed for these radicals from the protons, or from ring atoms such as ^{14}N or ^{13}C. Moreover, as was noted originally by Weissman[2] and by McConnell,[3] the a factors are roughly proportional to the π-electron spin densities in the ring calculated from simple Hückel theory. McConnell[3,4] was the first to express this relationship concisely for proton splittings in an aromatic hydrocarbon radical as follows:

$$a_H = Q\rho^\pi. \qquad (5.29)$$

This is known as the McConnell relationship. ρ^π is the π-electron spin density at the carbon atom linked to the hydrogen in question, and Q is a constant. If we take as an example the benzene anion, the unpaired electron must be equally distributed amongst the ring atoms hence $\rho^\pi = \frac{1}{6}$. The observed splitting of 3·75 gauss therefore gives $Q = 22·5$ gauss. The sign of a is not determined by this type of experiment but from other data one knows that Q is negative.

The reason that hyperfine splitting is observed for these radicals is that exchange forces between electrons couple the spins of the σ and π electrons so that a small spin density appears in the σ orbitals. The first quantitative interpretation of this was by McConnell,[4] Bersohn[5] and Weissman.[6] The first two authors developed the theory from valence-bond wave functions and the latter from molecular orbital functions. We will outline the MO approach.

As a first approximation to the wave function of the radical we have a set of electrons paired in σ orbitals, a set paired in π orbitals, and an unpaired electron in a π orbital. The Slater determinant for this function we represent by

$$\Psi = |\sigma_{1\alpha}\sigma_{1\beta} \cdots \sigma_{k\alpha}\sigma_{k\beta}\pi_{1\alpha}\pi_{1\beta} \cdots \pi_{m\alpha}\pi_{m\beta}\pi_{n\alpha}|. \tag{5.30}$$

Promoting an electron from an occupied σ orbital (σ_i) to a vacant (anti-bonding) σ orbital (σ_p) gives a configuration with three unpaired electrons, and from this it is possible to construct a quartet and two doublet spin states. The doublet state relevant to the problem we are concerned with is the one with the spin function

$$\sqrt{\tfrac{1}{6}}\{2\alpha\alpha\beta - \alpha\beta\alpha - \beta\alpha\alpha\}, \tag{5.31}$$

where we write, in order, the spins associated with σ_i, σ_p and π_n. The doublet-state wave function, written in full, consists of three determinants like (5.30) with appropriate spin functions: we refer to the resulting state as $\Psi_{i,p}$.

Under the operation of the electron repulsion terms in the Hamiltonian the wave function of the radical takes the form

$$\Psi' = \Psi + \sum_i \sum_p c_{ip}\Psi_{i,p} \tag{5.32}$$

where, by perturbation theory,

$$c_{ip} = -\langle\Psi|\sum_{i<j} e^2/r_{ij}|\Psi_{i,p}\rangle/\Delta E_{i,p}, \tag{5.33}$$

$\Delta E_{i,p}$ being the excitation energy required to promote the electron in the σ system. On evaluating the matrix element in (5.33) this reduces to

$$c_{ip} = \sqrt{\tfrac{3}{2}}(\sigma_i\pi_n \mid \pi_n\sigma_p)/\Delta E_{i,p}. \tag{5.34}$$

From the wave function (5.32) the leading contribution to the spin density in the σ system is calculated from the cross terms

$$2\sum_i \sum_p c_{ip}\langle\Psi| 2\sum_k S_k \, \delta(\mathbf{r}_{kN})|\Psi_{i,p}\rangle. \tag{5.35}$$

Developing this expression gives

$$\rho(r_N) = \frac{4}{\sqrt{6}}\sum_i \sum_p c_{ip}\int \sigma_i \, \delta(\mathbf{r}_{kN})\sigma_p \, dv, \tag{5.36}$$

which after insertion of (5.34), can be written

$$\rho(r_N) = 2 \sum_i \sum_p \frac{(\sigma_i \pi_n \mid \pi_n \sigma_p)}{\Delta E_{i,p}} \sigma_i(r_N)\sigma_p(r_N). \tag{5.37}$$

If the π molecular orbitals are expanded in LCAO form, (5.27), the two-electron exchange integral in (5.37) can be written

$$(\sigma_i \pi_n \mid \pi_n \sigma_p) = \sum_\mu \sum_\nu c_{n\mu} c_{n\nu} (\sigma_i \phi_\mu \mid \phi_\nu \sigma_p). \tag{5.38}$$

The product of atomic orbital coefficients now defines the spin density in the π system

$$\rho_{\mu\nu}{}^\pi = c_{n\mu} c_{n\nu} \tag{5.39}$$

hence expression (5.37) takes the form

$$\rho(r_N) = 2 \sum_i \sum_p \frac{\sigma_i(r_N)\sigma_p(r_N)}{\Delta E_{i,p}} \sum_\mu \sum_\nu (\sigma_i \phi_\mu \mid \phi_\nu \sigma_p)\rho_{\mu\nu}{}^\pi. \tag{5.40}$$

Thus the expression for the hyperfine constant is from (5.24)

$$a_N = \tfrac{16}{3}\pi g \beta g_N \beta_N \sum_i \sum_p \frac{\sigma_i(r_N)\sigma_p(r_N)}{\Delta E_{i \to p}} \sum_\mu \sum_\nu (\sigma_i \phi_\mu \mid \phi_\nu \sigma_p)\rho_{\mu\nu}{}^\pi \tag{5.41}$$

which can be written

$$a_N = \sum_\mu \sum_\nu Q_{\mu\nu} \rho_{\mu\nu}{}^\pi, \tag{5.42}$$

or more concisely

$$a_N = \text{trace } Q\rho^\pi, \tag{5.43}$$

where

$$Q_{\mu\nu} = \frac{16\pi}{3} g\beta g_N \beta_N \sum_i \sum_p \frac{\sigma_i(r_N)\sigma_p(r_N)}{\Delta E_{i,p}} (\sigma_i \phi_\mu \mid \phi_\nu \sigma_p). \tag{5.44}$$

A more general derivation of expression (5.43) has been given by McLachlan, Dearman and Lefebvre[7] which assumes only that in a first approximation the wave function of the radical is an antisymmetrized product of separable π and σ functions, but not necessarily a single determinant as is (5.30). This approach allows one to use more flexible wave functions to calculate the π spin densities.

If we take a basis of localized σ orbitals to calculate Q, from expression (5.44), then only those associated with the bonds involving atom N will have a non-zero density at that nucleus. For a proton this will be the σ orbitals of the CH bond. In that case the exchange integral $(\sigma_i \phi_\mu \mid \phi_\nu \sigma_p)$ will be very small unless both ϕ_μ and ϕ_ν are the π orbital of the carbon atom. In this approximation therefore, only one term effectively contributes to a_N and the simple McConnell formula (5.29) is valid.

The first interpretation of the hyperfine splittings of aromatic radicals was based on Hückel theory, and the results were quite encouraging. From the pairing property of the π orbitals of alternants one could understand that the a factors of corresponding positive and negative ions (e.g. anthracene cation and anion) were very similar. Carrington[8] has reviewed the results of Hückel theory.

Hückel calculations on neutral odd-alternant radicals like benzyl give zero spin densities at alternate positions in the conjugated chain. Nevertheless hyperfine splittings are observed from protons at such positions and their interpretation therefore needs a more sophisticated treatment of the π electrons. Hoijtink[9] and McConnell and Chesnut[10] showed that configuration interaction between the normal single determinant function (5.30), and excited configurations obtained by promoting electrons from the doubly occupied π orbitals to vacant π orbitals could explain this. McLachlan[11] showed the same thing by a simplified treatment of the unrestricted Hartree–Fock method. From these calculations it is found that the positions having zero spin density in the Hückel method, end up with negative spin densities. That is, for the $S_z = +\frac{1}{2}$ eigenstate there is more chance of finding a β electron at these positions than an α electron.

5.3 Calculation of π-electron spin densities of organic radicals

Table 5.1 shows the results of some calculations on two neutral aromatic radicals and on the azulene anion, with atoms numbered as in figure 5.4.

TABLE 5.1 Comparison of observed and calculated π-electron spin densities: the observed values were deduced from hyperfine constants on the assumption that $Q = -27$ gauss. Their signs have been assumed positive except where moduli are indicated. Bernal, Rieger and Fraenkel[14] have, for azulene, taken Q to vary with the ring angles, and their resulting spin densities differ slightly from those quoted here

		Hückel	McL[11,14]	UHF$^{(a),16}$	UHF$^{(b),16}$	RHF + CI[15]	Exp.[12,13,14]
Benzyl	1	0·143	0·161	0·157	0·205	0·125	0·190
	2	0	−0·063	−0·050	−0·105	−0·032	|0·065|
	3	0·143	0·137	0·128	0·175	0·086	0·227
	7	0·572	0·770	0·718	0·752	0·804	0·607
Perinaphthenyl	1	0	−0·070	−0·064	−0·133	−0·053	|0·067|
	2	0·167	0·226	0·219	0·277	0·215	0·233
Azulene anion	1	0·107	0·119	0·099	0·080	0·146	0·146
	2	0·005	−0·027	−0·006	−0·001	−0·011	|0·010|
	3	0·213	0·292	0·276	0·236	0·222	0·230
	4	0·011	−0·081	−0·115	−0·046	−0·051	|0·049|
	5	0·256	0·368	0·397	0·356	0·372	0·326

The table shows the results of restricted Hartree–Fock calculations followed by configuration interaction (RHF + CI),[15] of unrestricted Hartree–Fock calculations (UHF) before and after spin projection (*a* and *b* respectively),[16] and of the McLachlan method.[11] The latter is essentially a perturbation treatment of the UHF solution starting from the RHF wave function, in which the perturbation to the spin densities is related to the atom–atom polarizabilities defined in Hückel theory.

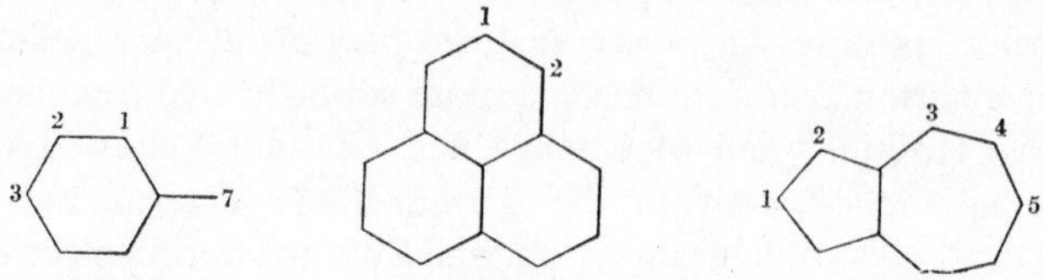

Benzyl Perinaphthenyl Azulene

FIGURE 5.4 Numbering of the atoms for the molecules referred to in table 5.1

In the UHF method the wave function is written as a single determinant but different orbitals are taken for the α and β electrons. Thus the wave function has the form

$$\Psi = |\psi_{1\alpha}\psi_{2\alpha}\cdots\psi_{m\alpha}\psi_{n\alpha}\psi_{1\beta}'\psi_{2\beta}'\cdots\psi_{m\beta}'|. \tag{5.45}$$

The LCAO–SCF equations for this wave function were first given by Pople and Nesbet[17] and their ZDO form was derived and used by Brickstock and Pople.[18] These equations are given in appendix 2.

The wave function (5.45) as written is not an eigenfunction of the total spin operator $\mathbf{S}^2$ according to the following equation

$$\mathbf{S}^2\Psi = s(s + 1)\Psi. \tag{5.46}$$

If there are $2n - 1$ electrons, then it is a mixture of doublet plus quartet plus higher states up to those of multiplicity $2n$. As the true wave function of the radical should be a pure doublet state ($s = \frac{1}{2}$, multiplicity $2s + 1 = 2$), one should properly derive an SCF wave function which is unrestricted, as (5.45), but has all states of higher multiplicity than 2 projected out. This spin projection is most easily done after the general UHF function has been derived. The technique is simply to apply in succession the operators

$$[\mathbf{S}^2 - s(s + 1)] \tag{5.47}$$

to Ψ, with $s = \frac{3}{2}, \frac{5}{2}, \ldots$, these will annihilate all component spin states except the doublet for which $s = \frac{1}{2}$. There is little to choose between the data shown in table 5.1 for all the calculations based on SCF orbitals.

The results for benzyl shown in table 5.1 are unsatisfactory in so far as they give a poor value for the ratio of the ortho and para spin densities. Recent calculations suggest that one needs to allow for unequal bond lengths in the molecule to get the correct ratio.[19] In a calculation which used the optimum scf bond lengths of the benzyl cation the results shown in figure 5.5 were obtained. It is seen that the ratio of ortho and para spin

scf bond lengths of the benzyl cation (Å)

ρ_π

a

FIGURE 5.5 scf bond lengths of the benzyl cation and the resulting spin densities of the benzyl radical. The optimum bond lengths of the benzyl anion are close to those of the cation

densities obtained from the self-consistent geometry is now close to one. The spin densities were calculated by the INDO method (§ 5.5).

There have been only a small number of scf calculations on the spin densities in hetero-atomic systems. The azine anion radicals have received some attention as they provide a good series of experimental data. Black and McDowell[20] have varied the hetero-atom parameters in these calculations and conclude that a wider spread of values is obtained by varying the parameters than by varying the method of calculation. Other calculations on these molecules have been made by Ali and Hinchliffe.[21]

The spin densities of the anion of benzene and substituted benzenes have received considerable attention. If we take the spin density in benzene for the hexagonal configuration to be an average $\frac{1}{6}$ at each atom then from the observed hyperfine splitting of 3·825 gauss[22] we deduce $Q = 23$,† which is appreciably smaller than the best Q for the larger aromatic radicals. The complication for the benzene anion is that it has an electronically degenerate ground state and would be subject to a Jahn–Teller distortion. Further evidence of this complication is that the deuterobenzene anion shows non-equivalent proton splittings.[23,25] Various attempts have been made to improve the benzene anion calculations, by configuration

† It is known that the spin density at an aromatic proton is negative so that a and Q should be negative. But the sign is often ignored as it is not determined by the standard ESR experiment. In table 5.2 results of direct calculations are given with the correct sign.

interaction,[24] by the unrestricted Hartree–Fock method[16] and by allowing for distortion,[26] but none of these explain the low value observed for Q.

The most likely explanation is that the Q factor depends on the charge located at the carbon atom. This would not only explain why Q for the benzene anion was smaller than that for the larger hydrocarbon radical anions, but also why Q for a radical cation is generally slightly larger than for the corresponding radical anion. The charge effect may be due to a general expansion of orbitals with increasing negative charge and a corresponding reduction in the value of the exchange integrals (5.38) which

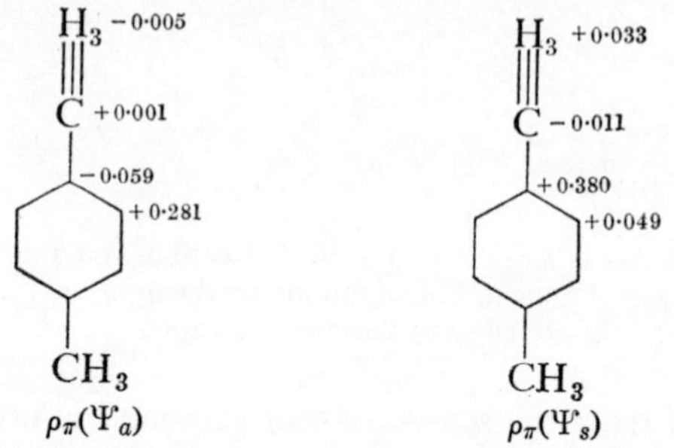

FIGURE 5.6   The effect of alkyl groups on the spin states of the benzene anion

enter the derivation of the McConnell relationship.[27] Alternatively there have been explanations based on a more extensive analysis of electron correlation.[28] Sales[29] has done a critical analysis of these theories.

Several calculations have been made on the radical anions of the alkyl benzenes. A single alkyl group, as in toluene, destabilizes the symmetric component of the benzene anion ground state relative to the antisymmetic component, so that the odd electron occupies the antisymmetric orbital and there is only a very small spin density on the alkyl group. This destabilization is introduced through an inductive perturbation of the ring. That is, the alkyl group makes the carbon atom to which it is attached less attractive to electrons. Nevertheless, the transfer of spin density to the alkyl protons appears to occur through a hyperconjugative (mesomeric) effect, so the situation is quite complicated. The results of de Boer and Colpa,[30] for the spin densities of the two states of p-xylene are shown in figure 5.6. As no temperature variation of the hyperfine splitting is observed for this radical the antisymmetric state Ψ_a must be appreciably lower in energy than the symmetric state.

Other series of substituted benzene radical anions that have been studied quite extensively are those derived from cyano and nitrobenzenes. References may be found in the article by Sales.[29]

5.4   The calculation of ESR Q factors

Using reasonable values for the exchange integrals and excitation energies that appear in (5.44), and the assumption of localized C—H wave functions, the first estimates of Q for a proton were in the region of 25 gauss. More reliable calculations have since been made on the Q factors for carbon,[31] nitrogen[32] and fluorine,[33] but in most cases the σ orbitals were based on independent electron calculations. For example, Hinchliffe and Murrell[33] calculated Q factors needed for the interpretation of fluorine couplings as follows:

$$Q_{FF} = +200, Q_{CF} = +5, Q_{FC} = -62, Q_{CC} = -11 \text{ gauss.}$$

There is no evidence that these are seriously wrong, but unfortunately it has not been found possible to deduce reliable empirical values for these constants by fitting calculated π spin densities to experimental splittings.

Melchior[34] has examined in detail the sensitivity of Q factors to assumptions regarding the σ-electron wave functions. He has applied a formulation based on orthogonalized σ orbitals to CH_3 and $C_2H_4^-$. Broze and Luz[35] have calculated the Q factors for the ^{17}O hyperfine splitting of a carbonyl group using a basis of SCF orbitals of formaldehyde derived by Newton and Palke.[36] Their results are

$$Q_{OO} = 48 \cdot 70, \quad Q_{CC} = -0 \cdot 46 \quad \text{and} \quad Q_{CO} + Q_{OC} = -6 \cdot 03 \text{ gauss.}$$

which have a similar balance to the values quoted above for fluorine.

Several calculations of Q factors have been made with the valence bond method, and these are reviewed by Sales.[29]

5.5   INDO calculations of spin densities

Because one-centre exchange integrals are retained in INDO calculations, an unrestricted Hartree–Fock calculation on a π-electron radical with this model will automatically give some spin density in the σ system. The hyperfine constants can therefore be calculated directly by this method without the need to use the π-electron spin density and Q factors.

Calculations were first made with this approach by Pople, Beveridge and Dobosh.[37] No attempt was made to project out the components of higher spin multiplicity according to the procedure of expression (5.47) A large number of compounds were studied and an analysis was made of the correlation between spin density in the valence s orbital of the relevant nucleus and the observed hyperfine constant. For protons a correlation coefficient of 0·88 was found from 141 data points. For ^{13}C, and ^{19}F the results were rather better, although with fewer data. For ^{14}N and particularly ^{17}O (only 5 data) the correlation was rather poor.

For the π-electron radicals the results paralleled those obtained from the McConnell relationship, a close proportionality being found between the hydrogen $1s$ spin density and the π spin density on the attached carbon atom. Table 5.2 gives a selection of results for proton hyperfine constants.

TABLE 5.2   Comparison of observed and calculated hyperfine constants obtained by the INDO method[37]

		Calc.	Obs. (sign assumed)
cyclopentadienyl (C_5H_5)		−4·8	−5·60
benzene anion ($C_6H_6{}^-$)		−3·6	−3·75
tropyl (C_7H_7)		−3·2	−3·95
anthracene anion			
	1	−2·7	−2·74
	2	−0·6	−1·51
	9	−6·8	−5·34
anthracene cation	1	−2·9	−3·08
	2	−0·6	−1·38
	9	−6·6	−6·49

Several other small and medium-sized radicals have since been studied by the INDO method. Thomson[38] found that the spin densities for the σ radicals HBO$^-$, HCO and HCN$^-$ agreed satisfactorily with experimental hyperfine splittings providing that the geometries were optimized. Danen and Kensler[39] for the π radical NMe$_2$ and Bakuzis and coworkers[40] for the σ radical norbornenyl, have deduced geometries for the radicals by fitting calculations to the hyperfine parameters.

Kasai and coworkers[41] investigated the aryl radicals (phenyl and naphthyl) to determine whether their ground states correspond to σ or π radicals. They were found to be σ in all cases but for the naphthyl radicals the highest occupied molecular orbital had π symmetry, that is the unpaired electron was found not to occupy the orbital of highest energy.

Krusic and Rettig[42] from calculations on the σ radical benzoyl, confirmed the rather surprising experimental observation that only the meta-ring protons showed a large hyperfine interaction.

These results show what is perhaps the outstanding success of the INDO method. No attempt has yet been made to apply spin projection and it is to be hoped that if this is done the correlation will not be poorer. The π-electron calculations listed in table 5.1 show that spin-projection can have an appreciable effect on spin densities.

5.6 Nuclear magnetic resonance

The analysis of high resolution NMR spectra leads to two quantities susceptible to theoretical interpretation, the screening constants or chemical shifts and the coupling constants. The first of these is defined by the relationship

$$H = H_0(1 - \sigma) \tag{5.48}$$

where H_0 is the external applied magnetic field, and H is the field actually experienced by the nucleus. For space fixed molecules σ is a tensor, because H depends on the orientation of the molecule relative to the direction of the applied field. However, from high resolution NMR we only measure the average value of this tensor as

$$\sigma = \tfrac{1}{3}(\sigma_{xx} + \sigma_{yy} + \sigma_{zz}). \tag{5.49}$$

The screening constant σ has its origin in the circulations of electrons which are stimulated by the applied field. These circulations have associated with them secondary magnetic fields which may either enhance or militate against H_0.

The Zeeman Hamiltonian for several nuclei, N, with a magnetic field in the negative z direction (the conventional choice in NMR), is

$$\sum_{N} g_N \beta_N \hbar H_0(1 - \sigma_N) I_{Nz}. \tag{5.50}$$

It is usual to divide this by h to have the eigenvalues directly in frequency units, and express it more simply as

$$\sum_{N} \nu_N I_{Nz} \tag{5.51}$$

where

$$\nu_N = g_N \beta_N (2\pi)^{-1} H_0(1 - \sigma_N). \tag{5.52}$$

This Hamiltonian has $2I + 1$ eigenvalues corresponding to the different values of the z component of $\mathbf{I}_N$ ($M_I = I, I - 1, \ldots, -I$). These are separated by equal intervals of ν_N. The magnetic dipole transitions are subject to the selection rule $\Delta M_I = \pm 1$ and hence all transitions occur at a frequency ν_N.

In addition to the Zeeman Hamiltonian there will be terms arising from the interaction of the nuclear spins with one another. The direct dipole–dipole interaction between nuclei averages to zero in a tumbling molecule (cf. 5.16). The fine structure in high resolution NMR spectra arises from the so-called indirect coupling terms which have the form of a scalar interaction between each pair of magnetic moments

$$\sum_{N > N'} J_{NN'} \mathbf{I}_N \cdot \mathbf{I}_{N'}. \tag{5.53}$$

J is called the coupling constant.

The complete spin Hamiltonian for the interpretation of high resolution NMR is therefore a combination of (5.51) and (5.53)

$$\mathcal{H} = \sum_N \nu_N I_{Nz} + \sum_{N > N'} J_{NN'} \mathbf{I}_N \cdot \mathbf{I}_{N'}. \tag{5.54}$$

It can be seen that this spin Hamiltonian is structurally rather similar to that used for ESR spectra (5.14). The mathematics is therefore similar in the two cases, the most important difference being that off-diagonal matrix elements of $\mathbf{I}_N \cdot \mathbf{I}_{N'}$ play an important part in the analysis of NMR spectra, whereas we saw that off-diagonal elements of $\mathbf{I} \cdot \mathbf{S}$ could be ignored in ESR.

The full matrix of $\mathcal{H}$ for a system of two spin-$\frac{1}{2}$ nuclei is given in table 5.3. The matrix elements are readily derived if one follows the mathematics of equations (5.8) to (5.14).

If $J/2 \ll \nu_A - \nu_B$ the off-diagonal matrix element may be neglected, and we have a so-called first-order or AX spectrum as shown in figure 5.7. If the above condition does not hold, then we have to diagonalize the 2×2 matrix to get the resulting eigenvalues and we have a second-order or AB spectrum.

The analysis is trivial for a two-spin system but can become quite tricky for larger numbers of nuclei. It is beyond the scope of this book to describe the techniques involved, and we refer the reader to a standard text,[43,44] or to the review by Bishop.[45]

The electronic theory of shielding constants and coupling constants was first outlined by Ramsey and Purcell.[46,47,48] Good accounts of this can be

TABLE 5.3   Matrix elements of the spin Hamiltonian $\mathcal{H}$ for two spin-$\frac{1}{2}$ nuclei. The basis functions define the spin states of the two nuclei (AB)

$\lvert\alpha\alpha\rangle$	$\frac{1}{2}(\nu_A + \nu_B) + J/4$	0	0	0
$\lvert\alpha\beta\rangle$	0	$\frac{1}{2}(\nu_A - \nu_B) - J/4$	$J/2$	0
$\lvert\beta\alpha\rangle$	0	$J/2$	$-\frac{1}{2}(\nu_A - \nu_B) - J/4$	0
$\lvert\beta\beta\rangle$	0	0	0	$-\frac{1}{2}(\nu_A + \nu_B) + J/4$

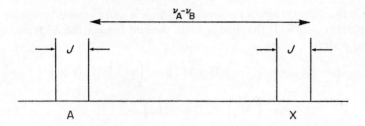

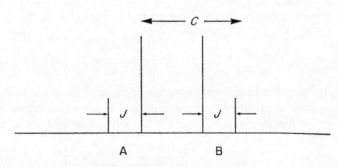

FIGURE 5.7 The AB and AX spectra for two spin-$\frac{1}{2}$ nuclei. For the AB case the separation C is given by

$$C = (J^2 + (\nu_B - \nu_A)^2)^{\frac{1}{2}}$$

obtained from many of the standard texts on magnetic resonance and from several reviews.[43,44,49-51] We will just outline the important steps in the theory.

5.7 NMR screening constants

In the presence of a magnetic field $\mathbf{H}$ an electron experiences an additional potential energy of the form

$$\frac{e}{mc}\mathbf{A} \cdot \mathbf{p} + \frac{e}{2m^2c^2}\mathbf{A}^2, \tag{5.55}$$

where $\mathbf{A}$ is the vector potential

$$\mathbf{A} = \tfrac{1}{2}\mathbf{H} \times \mathbf{r}, \tag{5.56}$$

and $\mathbf{p}$ is the momentum of the electron which is more explicitly written as the operator $-i\hbar\nabla$. If the field is small we can neglect the $\mathbf{A}^2$ term, and use the operator

$$\left(\frac{e}{mc}\right)\mathbf{A} \cdot \mathbf{p} = \left(\frac{e}{2mc}\right)(\mathbf{H} \times \mathbf{r}) \cdot \mathbf{p} = \left(\frac{e}{2mc}\right)[\mathbf{H} \cdot (\mathbf{r} \times \mathbf{p})]$$

$$= \left(\frac{e\hbar}{2mc}\right)\mathbf{H} \cdot \mathbf{L} = \beta\mathbf{H} \cdot \mathbf{L}. \tag{5.57}$$

In this derivation we have used the fact that $\mathbf{r} \times \mathbf{p}/\hbar$ is the orbital angular momentum and $e\hbar/2mc$ is the Bohr magneton. There will be a sum of such terms, one for each electron.

Using perturbation theory we can write the electronic wave function as

$$\Psi = \Psi_0 - \beta H \sum_n \frac{\langle n | \sum_k L_{kz} | 0 \rangle}{E_n - E_0} \tag{5.58}$$

where Ψ_0 is the ground state wave function in the absence of the field.

We now wish to calculate the magnetic field at a nucleus arising from the motion of the electrons. Classically we write this field for a single electron

$$-\mathbf{H}' = \left(\frac{e}{c}\right)\frac{\mathbf{r} \times \mathbf{v}}{r^3} = \left(\frac{e}{mc}\right)\frac{\mathbf{r} \times \left(\mathbf{p} + \dfrac{e}{c}\mathbf{A}\right)}{r^3}$$

$$= \frac{2\beta\mathbf{L}}{r^3} + \left(\frac{e^2}{2mc^2}\right)\frac{\mathbf{r} \times (\mathbf{H} \times \mathbf{r})}{r^3} \tag{5.59}$$

and in quantum mechanics we take the average of this operator over the state Ψ'. For simplicity we take only the z component of $\mathbf{H}'$

$$-H_z' = \langle\Psi| \sum_k 2\beta L_k/r_k^3 |\Psi\rangle + \left(\frac{e^2}{2mc^2}\right)H\langle\Psi| \sum_k (x_k^2 + y_k^2)/r_k^3 |\Psi\rangle. \tag{5.60}$$

This can be equated to the shielding term which from (5.58) is $\sigma_{zz}H_0$. After substituting the wave function Ψ from (5.48) we can equate the terms linear in H_0, and obtain the expression

$$\sigma_{zz} = \left(\frac{e^2}{2mc^2}\right)\langle 0| \sum_k (x_k^2 + y_k^2)/r_k^3 |0\rangle$$

$$- \beta^2 \sum_n \left[\langle 0| \sum_k L_{kz} |n\rangle\langle n| \sum_k 2L_{kz}/r_k^3 |0\rangle\right.$$

$$\left. + \langle 0| \sum_k 2L_{kz}/r_k^3 |n\rangle\langle n| \sum_k L_{kz} |0\rangle\right](E_n - E_0)^{-1}. \tag{5.61}$$

Equation (5.61) is called Ramsey's shielding formula. There are similar expressions for σ_{xx} and σ_{yy}, and from these one can evaluate the average isotropic screening constant σ. For real wave functions this becomes

$$\sigma = \frac{e^2}{3mc^2} \langle 0 | \sum_k r_k^{-1} | 0 \rangle - \frac{4\beta^2}{3} \sum_n \langle 0 | \sum_k \mathbf{L}_k | n \rangle \cdot \langle n | \sum_k \mathbf{L}_k / r_k^3 | 0 \rangle (E_n - E_0)^{-1}.$$

(5.62)

The first term in (5.62) is usually called the diamagnetic term by analogy with a corresponding term that occurs in the theory of magnetic susceptibility. Physically it corresponds to free rotation of the electrons in the molecule around the nucleus in question. However, such a rotation is hindered by the presence of the other nuclei in the molecule, and this is represented by the second term in (5.62), which is called the paramagnetic term.

There are two problems encountered in evaluating expression (5.62). The first is that it is based on a perturbation formula so that some knowledge of the excited states n is required. However, variational wave functions have been used to replace the sum over all excited states. Secondly, for a large molecule both the diamagnetic and paramagnetic terms are very large and nearly cancel. For this reason most calculations have assumed, following Saika and Slichter,[52] that one can separate local circulations around the individual atoms from the overall motion of the electrons in the molecule. This has been developed in a molecular-orbital framework by Pople,[53] and we present a brief description of his theory.

The total shielding constant for a nucleus A is written

$$\sigma_A = \sigma_d{}^{AA} + \sigma_p{}^{AA} + \sum_{B \neq A} \sigma^{AB} + \sigma^{A,ring}$$

(5.63)

where the following interpretation is given to these separate terms.

1. $\sigma_d{}^{AA}$ is the diamagnetic contribution from circulations around the atom in question. For an isolated spherical atom it is given by the first term in (5.62) which is known as the Lamb formula.[54] To calculate it one only needs to know the average value of r^{-1} for the atomic wave function. For the molecule this term appears as a contribution from each atomic orbital weighted by the population of that orbital.

$$\sigma_d{}^{AA} = \frac{e^2}{3mc^2} \sum_\mu P_{\mu\mu} (r^{-1})_{\mu\mu}.$$

(5.64)

2. $\sigma_p{}^{AA}$ represents the effect of local paramagnetic currents. Saika and Slichter[52] showed that these were more sensitive to molecular structure

than $\sigma_d{}^{AA}$. For example, the difference of $\sigma_p{}^{AA}$ for fluorine in F_2 and in an ionic fluoride is a factor of 10^{-3}. It can be seen from (5.62) that the term arises from the mixing of electronic states by the magnetic field. States of the appropriate symmetry involve the occupation of atomic p orbitals. Thus for hydrogen which has no low-lying p orbitals, the paramagnetic term is less important than it is for an atom like fluorine. In most work the difficulty in evaluating the summation over n in (5.62) is circumvented by the average-energy approximation. By replacing $E_n - E_0$ by a constant ΔE, considered as the average excitation energy, the sum over states is reduced to

$$\frac{4\beta^2}{3} \langle 0| \sum_j \sum_k \mathbf{L}_j \cdot \mathbf{L}_k r_k^{-3} |0\rangle / \Delta E. \tag{5.65}$$

Unfortunately this approximation does not lead to an expression for σ which is gauge invariant, that is, whose value is independent of the origin of coordinates. An alternative averaging procedure which maintains invariance has been proposed by Kern and Lipscomb[55] and Sadlej.[56]

Pople[53] has derived a molecular orbital expression for (5.65) which involves bond orders between the p orbitals of the atom in question and its neighbours (B)

$$\sigma_p{}^{AA} = -\frac{2e^2\hbar^2}{3m^2c^2\,\Delta E} \langle r^{-3}\rangle_p \sum_B \{\delta_{AB}(P_{x_A x_B} + P_{y_A y_B} + P_{z_A z_B})$$
$$- \tfrac{1}{2}(P_{y_A y_B}P_{z_A z_B} + P_{z_A z_B}P_{x_A x_B} + P_{x_A x_B}P_{y_A y_B})$$
$$+ \tfrac{1}{2}(P_{y_A z_B}P_{z_A y_B} + P_{z_A x_B}P_{x_A z_B} + P_{x_A y_B}P_{y_A x_B})\}. \tag{5.66}$$

3. If the contribution from electron currents on neighbouring atoms is approximated by a dipolar field we can write

$$\sigma^{AB} = -\sum_{\alpha\beta} \chi_{\alpha\beta}{}^B (3R_{B\alpha}R_{B\beta} - \delta_{\alpha\beta}R_B{}^2)/3NR_B{}^5, \tag{5.67}$$

α and β being the tensor suffixes to run over the x, y and z components. R_B is the vector position of atom B relative to A as origin. The elements of χ^B the *molar* magnetic susceptibility are given by combining a paramagnetic term given by an expression similar to (5.66), and a diamagnetic term

$$(\chi_d{}^B)_{zz} = -\frac{e^2}{4mc^2} \sum_\mu P_{\mu\mu}(x^2 + y^2)_{\mu\mu}. \tag{5.68}$$

4. The final term in (5.63) is special for molecules which can maintain electron currents around rings of atoms. Such currents are characteristic of planar aromatic molecules and arise from the delocalized π electrons.

The situation shown in figure 5.8 shows that such a current deshields the protons in benzene. That is, for a fixed frequency the resonance condition is fulfilled by smaller external magnetic fields than would be required if this current were absent.

For aromatic hydrocarbons simple estimates of the shielding due to the ring current can be made from the assumption that the π electrons are free to move on rings whose radii are determined by the carbon skeleton.

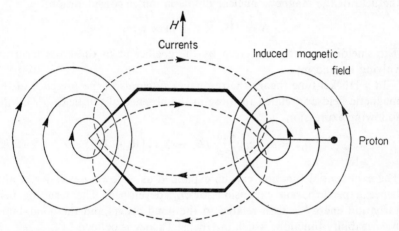

FIGURE 5.8   The induced magnetic field from the ring current of benzene[1]

The chemical shifts are then calculated by evaluating the magnetic field arising from these currents either in a dipolar approximation[57] or by calculating them exactly.[58,59] This is, however, an idealized model, and one should properly allow for a resistance to the electron circulation arising from the nuclear potentials. This is particularly important if there are heteroatoms in the rings. A molecular orbital theory of the ring current has been developed and its results will be described before we turn to the other contributions.[60,61]

If the molecular orbitals in the absence of a field are constructed from a set of atomic orbitals $\varphi_1{}^0 \cdots \varphi_m{}^0$ then in the presence of a field it was shown by London[62] that the most convenient basis of functions to use is

$$\varphi_k = \varphi_k{}^0 \exp\left(-\frac{ie}{\hbar c} \mathbf{A}_k \cdot \mathbf{r}\right), \tag{5.69}$$

where $\mathbf{A}_k$ is the magnetic vector potential at the kth nucleus, which is given by an expression of the form of (5.56). The phase factor is introduced here so that the matrix elements of the kinetic energy operator, which in the presence of a field is

$$-\sum_k \frac{\hbar^2}{2m}\left(\nabla_k + \frac{ie}{\hbar c}\,\mathbf{A}_k\right)^2, \tag{5.70}$$

shall not depend on the origin of the vector potential $\mathbf{A}_k$.

The total energy can be developed as a perturbation expansion in $\mathbf{A}$, the total magnetic field being a combination of an external field $\mathbf{H}^0$ and the field of the magnetic nuclear dipole $\mathbf{m}$ under consideration,

$$\mathbf{A} = \tfrac{1}{2}\mathbf{H}_0 \times \mathbf{r} + \mathbf{m} \times \mathbf{r}/r^3. \tag{5.71}$$

The shielding constant is given by the coefficient of the cross term involving $\mathbf{H}_0$ and $\mathbf{m}$.

In a Hückel-type theory the resonance integrals in the presence of the magnetic field are related to those in the absence of the field, β^0, by the following expression

$$\beta_{ij} = \beta_{ij}{}^0 \exp\left\{\frac{ie}{2\hbar c}\,(\mathbf{A}_i - \mathbf{A}_j)\,.\,(\mathbf{R}_i + \mathbf{R}_j)\right\}. \tag{5.72}$$

The exponential function can be expanded as a power series in $\mathbf{A}$, and hence a perturbation expansion for $\beta_{ij}{}^0$ developed. The resulting perturbation energy is then related to the bond orders and the bond–bond polarizability function, which in Hückel Theory is defined by

$$\pi_{\mathrm{rs,tu}} = \frac{\partial P_{\mathrm{rs}}}{\partial \beta_{\mathrm{tu}}} = \frac{\partial^2 E}{\partial \beta_{\mathrm{rs}}\,\partial \beta_{\mathrm{tu}}}\,, \tag{5.73}$$

E being the total energy. $\pi_{\mathrm{rs,tu}}$ is given by a simple expression involving the coefficients and energies of the unperturbed orbitals. The resulting expression for the ring current shielding is given in ref. 61.

In SCF theory the perturbation due to the magnetic field also has its origin in changes in the one-electron integrals, but the F-matrix elements are perturbed not only by changes in β_{ij} but also by the resulting changes in the bond-order matrix. The SCF perturbation expression for the shielding constants has been derived by Hall and Hardisson and is rather complicated.[63] Table 5.4 shows their results and compares them with those obtained from Hückel theory.

The results for sym-triazine were obtained with three different values of an electronegativity parameter for the nitrogen atom, and show that in the SCF calculation it is possible for a hetero atom to enhance the ring current. However, for this molecule the chemical shift will be dominated by other terms in (5.63) so no direct comparison with experiment is possible.

TABLE 5.4 Comparison of ring current chemical shifts by the Hückel and SCF methods.[64,65] Results quoted give the shielding effect relative to the protons of benzene

		Hückel	SCF	Obs[a]
Naphthalene	1	1·31	1·28	1·34
	2	1·17	1·15	1·12
Azulene	1	1·27	1·37	1·04
	2	1·14	1·23	1·37
	4	1·34	1·32	1·61
	5	1·22	1·18	0·82
	6	1·20	1·16	1·11
sym-triazine	$\delta_N = 0\cdot2$	0·98	1·04	
	0·5	0·90	0·89	
	0·8	0·76	0·67	

[a] Based on an experimental estimate of 1·55 ppm for the ring current shift in benzene.

It has been generally recognised that with the exception of protons the variation in shielding constant of a nucleus for a related series of molecules is dominated by the variation in the paramagnetic term in the Ramsey formula (5.62). Expression (5.65) which was obtained after applying the closure approximation, has been moderately successful in explaining ^{13}C shielding constants, particularly for a series of closely related compounds.[66] For atoms which have non-bonding orbitals, notably oxygen and nitrogen, the average-energy approximation is less satisfactory because the paramagnetic term may be dominated by one low-energy excitation. At the other extreme therefore, shielding constants have been correlated with the energy of the $n - \pi^*$ transitions[67] which is frequently associated with the lowest energy absorption band of these molecules.

Velenik and Lynden-Bell[68] have evaluated both terms in the Ramsey expression by independent-electron MO theory, without making the separation of (5.63) and without the average-energy approximation. With a judicious choice of the Hückel parameters they were able to get reasonable correlations between theory and experiment for nitrogen, oxygen and carbon chemical shifts. However, different correlation lines were obtained for linear, planar and tetrahedral carbon atoms as the theory did not give absolute values of the shielding constants. The reason for this was attributed to the adoption of constant Hückel coulomb intergrals

(α) in the calculations, no variation with the state of hybridization or with the nature of neighbouring atoms being allowed. It would be anticipated that these difficulties would disappear in an SCF framework, but we have been told that CNDO/2 calculations led to no improvement over the Hückel results.[69]

Davies[70] calculated ^{19}F shielding constants for the fluorobenzenes using CNDO wave functions and the average energy approximation. The results were rather unsatisfactory in that the best results were obtained with negative values of the average excitation energy ΔE.

For diatomic molecules Kern and Lipscomb[55] evaluated shielding constants from their gauge-invariant average-energy approximation already referred to. Good results were obtained in particular for protons. Sadlej[56] reports, without giving details, that the method also gives good results for the proton shielding of methane using CNDO wave functions. Ditchfield, Miller and Pople[71] have calculated diamagnetic and paramagnetic contributions to the shielding of some diatomic molecules using their finite perturbation method which will be described in the next section.

5.8   NMR coupling constants

In high-resolution NMR the coupling between nuclei arises from an indirect effect transmitted by the electrons. The direct dipole–dipole interaction of two magnetic nuclei is important in solid state NMR but averages to zero for molecules tumbling in solution.

Electrons can transmit a nuclear spin interaction either by their orbital motion or by their spins. We have already shown in the ESR section (5.2) that the spin interaction is divided into two parts; there is the long range dipolar interaction of electron and nuclear moments (5.15) and the Fermi contact interaction (5.18). Thus the theory of the coupling constant involves three electron-nuclear coupling terms, and it can be shown that they make separate contributions to the coupling constant. For coupling between two atoms, one or both of which is hydrogen, the Fermi contact term is by far the most important of the three. This is because p orbitals need to be invoked to bring in the orbital and spin-dipolar contributions. As most calculations have been carried out for proton couplings we will outline only the derivation of the Fermi-contact contribution. The detailed theory of all three contributions is available in standard texts and in reviews.[43,44,49,50]

The Fermi-contact operator (5.18) is a linear function of the nuclear spin operators $\mathbf{I}_N$ whereas the coupling term in the spin Hamiltonian (5.54) is bilinear. It follows that if perturbation theory is to be used the coupling constant will arise from the second-order energy associated with the Fermi-contact operator. The expressions we shall derive follow from such an analysis, although variation methods have also been used to calculate J.[51]

If we use the familiar Rayleigh–Schrödinger (sum-over-excited-states) form of perturbation theory, the second-order energy is

$$E'' = \left(\frac{8\pi g\beta\hbar}{3}\right)^2 \sum_n \sum_N \sum_{N'} \gamma_N \gamma_{N'} \langle 0| \sum_k \delta(\mathbf{r}_{kN})\mathbf{S}_k \cdot \mathbf{I}_N |n\rangle$$
$$\times \langle n| \sum_{k'} \delta(\mathbf{r}_{k'N'})\mathbf{S}_{k'} \cdot \mathbf{I}_{N'} |0\rangle (E_0 - E_n)^{-1} \quad (5.74)$$

where we have introduced the magnetogyric ratios defined by

$$\gamma_N = g_N \beta_N / \hbar \quad (5.75)$$

in accord with common practice.

The matrix elements involve integration of the perturbation operator over the electronic coordinates alone. The nuclear spin operators can, therefore, be taken outside the sum over excited states n. However, the resulting expression will not yet be of the form we require as there will be terms involving products $I_{N x} I_{N y}$ which do not occur in the nuclear spin Hamiltonian (5.54). We convert to the required form by taking the rotational average of (5.74),

$$\bar{U} = \tfrac{1}{3}(U_{xx} + U_{yy} + U_{zz}) \quad (5.76)$$

and by equating the coefficient of $\mathbf{I}_N \cdot \mathbf{I}_{N'}$ to J/h (the spin Hamiltonian has eigenvalues in frequency units) we obtain

$$J_{NN'} = \frac{2}{3h} \left(\frac{8\pi g\beta\hbar}{3}\right)^2 \gamma_N \gamma_{N'} \sum_n \langle 0| \sum_k \delta(\mathbf{r}_{kN})\mathbf{S}_k |n\rangle$$
$$\times \langle n| \sum_{k'} \delta(\mathbf{r}_{k'N'})\mathbf{S}_{k'} |0\rangle (E_0 - E_n)^{-1}. \quad (5.77)$$

For different isotopes of the same element the coupling constants differ only through the different magnetogyric ratios γ. It is therefore convenient to introduce a reduced coupling constant K, which depends only on the electronic environment of the nucleus:[72]

$$K_{NN'} = 2\pi J_{NN'} / \hbar \gamma_N \gamma_{N'}. \quad (5.78)$$

Taking also the value for the g-factor of the electron as 2, we have, from (5.77)

$$K_{NN'} = \frac{512\pi^2\beta^2}{27} \sum_n \langle 0| \sum_k \delta(\mathbf{r}_{kN})\mathbf{S}_k |n\rangle \cdot \langle n| \sum_{k'} \delta(\mathbf{r}_{k'N'})\mathbf{S}_{k'} |0\rangle (E_0 - E_n)^{-1}.$$

$$(5.79)$$

The operator $\sum_k \delta(\mathbf{r}_{kN})\mathbf{S}_k$ mixes a singlet ground state $|0\rangle$ with excited states $|n\rangle$ that are triplet spin states. Molecular orbital theory has been applied to the calculation of J at several levels of approximation. These depend on the method by which the molecular orbitals are calculated and the way in which these are used to construct the ground and excited states.

If the ground state is written as a single determinant of molecular orbitals, and the excited states are constructed by promoting electrons from occupied molecular orbitals ψ_l, to vacant (virtual) orbitals ψ_r, then the coupling constant is given by the expression

$$K_{NN'} = -\frac{256\pi^2\beta^2}{9} \sum_l \sum_r \psi_l(N)\psi_r(N)\psi_l(N')\psi_r(N')/^3\Delta E_{l,r} \quad (5.80)$$

where $\psi_l(N)\psi_r(N)$ is the value of the electron density $\psi_l\psi_r$ at nucleus N.

If the molecular orbitals are written in LCAO form the densities at the nuclei can be related to atomic orbital densities. In a ZDO approximation one would retain only the one-centre term φ_μ^2, and if there is only one valence s orbital per atom, we can write

$$K_{NN'} = -\frac{256\pi^2\beta^2}{9} s_N^2(0)s_{N'}^2(0) \sum_l \sum_r c_{lN}c_{rN}c_{lN'}c_{rN'}/^3\Delta E_{l,r} \quad (5.81)$$

where $s_N^2(0)$ is the s orbital density at the nucleus, and c_{lN} is the coefficient of orbital s_N in ψ_l.

If an independent-electron molecular orbital theory is used the triplet excitation energy is given by the difference in orbital energies. In this case the summation may be equated to a function $\pi_{NN'}$ which is known in Hückel theory as the atom–atom polarizability (c.f. 5.73),

$$\pi_{NN'} = \frac{\partial^2 E}{\partial \alpha_N \partial \alpha_{N'}} = -4 \sum_l \sum_r c_{lN}c_{rN}c_{lN'}c_{rN'}/(\epsilon_r - \epsilon_l) \quad (5.82)$$

and on substituting this into (5.81) we get the Pople–Santry expression for the coupling constant[72]

$$K_{NN'} = \frac{64\pi^2\beta^2}{9} s_N^2(0)s_{N'}^2(0)\pi_{NN'}. \quad (5.83)$$

Expressions (5.80–5.83) are the basis of most molecular orbital theories of spin–spin coupling in which at least one of the atoms is hydrogen. For completeness we give the two contributions which involve p orbitals of the atoms. The orbital contribution involves a summation over singlet excited states and in LCAO form the analogous expression to (5.81) is

$$K_{NN'}(\text{orbital}) = -\frac{16\beta^2}{3} \sum_{l} \sum_{r} \sum_{\lambda\mu\nu\sigma} c_{l\lambda}c_{r\mu}c_{l\nu}c_{r\sigma}\langle\varphi_\lambda|\, \mathbf{L}_N/r_N{}^3\, |\varphi_\mu\rangle$$

$$\times \langle\varphi_\nu|\, \mathbf{L}_{N'}/r_{N'}{}^3\, |\varphi_\sigma\rangle/^1\Delta E_{l,r}. \quad (5.84)$$

The spin-dipolar contribution involves excitation to triplet excited states and is

$$K_{NN'}(\text{dipolar}) = -\frac{4\beta^2}{3} \sum_{l} \sum_{r} \sum_{\lambda\mu\nu\sigma} c_{l\lambda}c_{r\mu}c_{l\nu}c_{r\sigma}\langle\varphi_\lambda|\, (3r_{N\alpha}r_{N\beta} - \delta_{\alpha\beta}r_N{}^2)/r_N{}^5\, |\varphi_\mu\rangle$$

$$\times \langle\varphi_\nu|\, (3r_{N'\alpha}r_{N'\beta} - \delta_{\alpha\beta}r_{N'}{}^2)/r_{N'}{}^5\, |\varphi_\sigma\rangle/^3\Delta E_{l,r} \quad (5.85)$$

the tensor suffixes representing a sum over x, y and z components.

Calculations of the coupling constants of hydrocarbons have been carried out using both the Pople–Santry[73] and Hoffmann[74] forms of independent-electron theory. The results have been reviewed[50] and we will not give detailed references here.

Table 5.5 shows the calculated H–H, H–C and C–C coupling constants obtained from CNDO/2 and $_i$NDO wave functions.[75,76] The integral parameters used had the standard values proposed by Pople and coworkers and were in no way chosen to get the best results for coupling constants. However, the density at the nucleus of the carbon $2s$ orbital was treated as a parameter, and as the C—H and C—C coupling constants are sensitive to this it is an important feature of the calculations. The optimum value was found to be 4·54 a.u. which is considerably higher than the density of a SCF $2s$ orbital of a free carbon atom (2·77 a.u.).

The calculations were based on expression (5.81) where the triplet states were chosen with no configuration interaction, and on the appropriate development of expression (5.79) after allowing for configuration interaction amongst the triplet states. No calculations were made with configuration interaction for the singlet ground state and the high value of the $2s$ density which parameterizes these calculations is probably associated with this deficiency. It is well known that molecular orbital theory without configuration interaction does not give sufficient electron correlation for the normal electron-pair bond.

The results obtained from these calculations were considerably better than those of the independent-electron model. It is interesting to compare the CNDO and INDO calculations particularly for geminal coupling constants as these are thought to be strongly influenced by one-centre exchange integrals. The INDO results are the better of the two, the largest improvement being for CH_4. However, for this case the results are not as good as those of Loéve and Salem[77] who used non-empirical SCF orbitals. Non-empirical SCF orbitals have also been used as a basis for C_2H_4 but the ZDO results given in Table 5.5 are not in this case significantly poorer.[78]

An alternative approach to the treatment of the Fermi-contact perturbation has been used by Pople, McIver and Ostlund.[79] Instead of using the Rayleigh–Schrodinger method they evaluate SCF orbitals directly for the Hamiltonian of the free molecule plus the Fermi-contact operator. A suitable scaling factor is used to multiply the Fermi-contact operator so that the energy shift that it produces is beyond the rounding error of the calculation. The technique, which is an extension of one first used to calculate electric polarizabilities by Cohen and Roothaan,[80] is called 'the

TABLE 5.5 Calculated reduced coupling constants for simple hydrocarbons (units $10^{20}\ cm^{-3}$), with and without configuration interaction[75,76]

	CNDO/2 no Cl	CNDO/2 + Cl	INDO no Cl	INDO + Cl	Obs.
H–H coupling constants					
CH_4 gem.	0·12	0·14	—	−0·15	−1·00
C_2H_6 trans	1·42	1·72	—	2·01	—
gauche	0·22	0·29	—	0·31	—
average	0·59	0·77	—	0·88	0·67
C_2H_4 trans	1·45	2·05	1·96	2·25	1·59
cis	0·75	0·95	0·75	0·95	0·98
gem.	0·61	0·75	0·65	0·63	0·21
C_2H_2	0·70	0·77	0·93	0·96	0·79
C–H (directly bonded)					
CH_4	35·1	41·0	42·7	45·2	41·8
C_2H_6	31·6	41·6	34·8	45·7	41·8
C_2H_4	43·8	54·6	46·4	57·1	52·3
C_2H_2	76·3	93·7	88·4	93·2	83·1
C–C (directly bonded)					
C_2H_6	26·3	34·3	26·6	34·8	45·6
C_2H_4	80·6	85·1	86·6	90·8	89·0
C_2H_2	211·3	216·5	220·5	219·9	225·9
C–H (not directly bonded)					
C_2H_6	0·00	−0·40	0·08	−0·96	−1·50
C_2H_4	0·48	−0·21	0·72	−0·88	−0·80
C_2H_2	2·66	7·05	5·03	10·24	16·40

TABLE 5.6 Calculated reduced coupling constants by the finite perturbation method[79]
(K in units of 10^{20} cm^{-3})

	CNDO/2	INDO	Obs.
H–H coupling constants			
CH$_4$ gem.	0·10	−0·51	−1·00
C$_2$H$_6$ trans	1·28	1·55	—
gauche	0·20	0·27	—
average	0·49	0·70	0·67
C$_2$H$_4$ trans	1·62	1·93	1·59
cis	0·74	0·78	0·98
gem.	0·71	0·27	0·21
C$_2$H$_2$	0·55	0·92	0·79
C–H (directly bonded)			
CH$_4$	30·9	40·7	41·8
C$_2$H$_6$	30·9	40·4	41·8
C$_2$H$_4$	42·3	51·9	52·3
C$_2$H$_2$	68·0	77·0	83·1
C–C (directly bonded)			
C$_2$H$_6$	—	55·1	45·6
C$_2$H$_4$	—	108·2	89·0
C$_2$H$_2$	—	215·7	225·9
C–H (not directly bonded)			
C$_2$H$_6$	−0·85	−2·38	−1·50
C$_2$H$_4$	−1·27	−3·83	−0·80
C$_2$H$_2$	1·82	0·83	16·40

finite perturbation method'. The coupling constant $J_{NN'}$ can be calculated from the spin density induced at N' by introducing the Fermi-contact operator for nucleus N.

Because the Fermi-contact operator mixes singlet and triplet spin states of the unperturbed molecule, the SCF calculation must be carried out as an open-shell procedure. Pople and coworkers use the method of different orbitals for different spins (unrestricted Hartree–Fock), described in section 5.3, but make no attempt to project out high-multiplicity spin components. A comparison has been made of the finite perturbation and the sum-over-states methods and it has been shown that they are not equivalent.[81]

Table 5.6 shows some of the results obtained by the finite perturbation method. Both the carbon $2s$ density and the hydrogen $1s$ density at the nucleus were treated as parameters in these calculations and given the values $2s_c^2(0) = 4·0318$ a.u., $1s_h^2(0) = 0·3724$ a.u. Similar conclusions may be drawn from the finite perturbation results and the standard perturbation results shown in table 5.5.

Coupling constants have also been calculated for fluorine nuclei with the finite perturbation method but the results were disappointing.[79] Blizzard and Santry[82] have extended the theory to include the orbital and dipolar perturbations and get greatly improved results. They find, for example, that the trend of J_{CF} in the series $CH_{4-n}F_n$, which is a decrease to CHF_3 and then an increase to CF_4, is not explained by the contact term alone, which increases steadily with n. They found that the orbital contribution is larger than the dipolar, in agreement with independent-electron calculations of Murrell, Stevenson and Jones.[83]

The rather satisfactory interpretation of coupling constants given by the semi-empirical SCF method must be viewed with caution. For example, no one has to this date carried out a non-empirical calculation of the coupling constant of HD to better than 75% of the experimental value.[84,85]† In addition, minimum basis set calculations, whether empirical or non-empirical give a negative coupling constant for HF, whereas the experimental evidence suggests that it is positive.[86,87] A positive value has been obtained only from a highly expanded gaussian atomic orbital basis.[86] Because the coupling constant only involves part (the cross term) of a second-order perturbation energy one cannot easily obtain bounds from a variational calculation, and hence it is possible to get an exact agreement with experiment from a poor variational function.[85,88] However, the results of tables 5.5 and 5.6 show that the empirical SCF methods are successful in correlating certain trends in coupling constants. They give some confidence that we have a basic understanding of the mechanism of nuclear spin–spin coupling.

5.9   Nuclear quadrupole coupling constants

Nuclei with spin quantum numbers greater or equal to unity ($I \geqslant 1$) have non-spherical charge distributions and as a result they possess electric quadrupole moments. A quadrupole moment is defined as the second moment of a charge distribution and therefore has, in general, nine non-zero components which are written as the elements of a second rank tensor. There is unfortunately more than one way of defining the components of a quadrupole moment; Buckingham[89] has reviewed the various possibilities. For nuclei the definition usually adopted is

$$Q_{ij} = \int \rho(3x_i x_j - \delta_{ij} r^2) \, dv \qquad (5.86)$$

† C. M. Dutta, N. C. Dutta and T. P. Das, *Phys. Rev. Letters* **25**, 1695 (1970), have recently reported a calculation using the complete set of states of H_2^+ which leads to a value of J_{HD} in good agreement with experiment.

where x_i is one of the three cartesian axes and ρ is the charge density per unit volume. As a nucleus is considered to be made up of discrete protonic charges and uncharged neutrons, (5.86) may be replaced by the summation over protons

$$Q_{ij} = \sum_p e(3x_{ip}x_{jp} - \delta_{ij}r_p^2).$$ (5.87)

It is always possible to choose axes such that all off-diagonal elements of the tensor (Q_{xy} etc.) are zero. We also note from the definition (5.86) that the tensor is traceless,

$$Q_{xx} + Q_{yy} + Q_{zz} = 0.$$ (5.88)

For nuclei the axis of spin is necessarily an axis of cylindrical symmetry. If this is labelled the z-axis, then

$$Q_{xx} = Q_{yy},$$ (5.89)

and hence, from (5.88)

$$Q_{zz} = -2Q_{xx}.$$ (5.90)

Thus the nuclear quadrupole moment is fully determined by a scalar quantity Q, which is usually expressed in terms of electron-charge units as follows

$$eQ = Q_{zz} = \sum_p e(3z_p^2 - r_p^2).$$ (5.91)

A nucleus may be elongated along the spin axis, like a cigar, in which case Q is positive, or it may be flattened like a discus in which case Q is negative. Q has the dimensions of length squared and since nuclei have radii approximately 10^{-12} cm, Q will be of the order of 10^{-24} cm^2.

Nuclear quadrupole moments have two important effects in molecular spectroscopy. Firstly, the quadrupole has an interaction energy with any electric field gradient at the nucleus arising from the overall electron and nuclear distribution in the molecule. Quadrupole resonance spectroscopy is a direct result of this. Secondly, because the electric quadrupole and magnetic dipole moments of the nucleus are both linked to the spin axis, the nuclear magnetic moments of quadrupolar nuclei are strongly coupled to the molecular axes via the interaction of Q with electric field gradients in the molecule. This coupling is responsible for the relaxation of quadrupolar nuclei in NMR spectroscopy. A detailed discussion of the influence of nuclear quadrupole moments on molecular spectroscopy may be found in books by Das and Hahn[90] and Abragam.[91] In this section we present an outline of the theory of the interaction of the quadrupole moment with internal electric fields of the molecule, and describe the information which can be deduced from this about molecular wave functions.

An electric quadrupole moment has an interaction energy with a non-uniform electric field of the form

$$E_Q = \tfrac{1}{6} \sum_{i,j} \left(\frac{\partial^2 V}{\partial x_i \, \partial x_j} \right) Q_{ij}, \tag{5.92}$$

where

$$\frac{\partial^2 V}{\partial x_i \, \partial x_j} = \frac{\partial \mathscr{E}_i}{\partial x_j} = \frac{\partial \mathscr{E}_j}{\partial x_i} \tag{5.93}$$

are the components of the electrostatic field gradient. Thus the field gradient, like Q, is represented by the components of a second rank tensor, and we can likewise choose principal axes such that it is diagonal. By Laplace's equation

$$\frac{\partial^2 V}{\partial x^2} + \frac{\partial^2 V}{\partial y^2} + \frac{\partial^2 V}{\partial z^2} = 0, \tag{5.94}$$

so that there are only two independent components of the tensor. These are normally expressed by the quantities

$$eq = \frac{\partial^2 V}{\partial z^2} \quad \text{and} \quad \eta = \left| \frac{(\partial^2 V/\partial x^2) - (\partial^2 V/\partial y^2)}{(\partial^2 V/\partial z^2)} \right| \tag{5.95}$$

where, by convention, the axes are defined by the condition $V_{zz} > V_{xx} \geqslant V_{yy}$. For the field arising from an axially symmetric electron distribution, as for example that in a diatomic molecule, the asymmetry parameter η will be zero.

The interaction energy (5.92) may be written as the inner product of the two tensors

$$E_Q = \tfrac{1}{6} \mathbf{V}_{ij} \cdot \mathbf{Q}_{ij}. \tag{5.96}$$

This will depend not only on the scalar quantities Q, q and η, but also on the relative orientation of the axis systems of the two tensors. To find the allowed relative orientations we need to use the appropriate quantum-mechanical operator corresponding to the classical energy (5.96).

The only operator which defines the orientation of the nucleus is the spin operator $\mathbf{I}$, so that the quadrupole moment must be written in terms of its components. The appropriate operator must be symmetric in the components $\mathbf{I}_i$ and $\mathbf{I}_j$ as these do not commute and, by comparison with (5.87), it must therefore be proportional to

$$\tfrac{3}{2}(\mathbf{I}_i \mathbf{I}_j + \mathbf{I}_j \mathbf{I}_i) - \delta_{ij} \mathbf{I}^2. \tag{5.97}$$

More precisely, we can use the Wigner–Eckart theorem to write a correspondence between matrix elements of the spin states $|I, M_I\rangle$ as follows

$$\langle I, M_I | \sum_p (3x_{ip}x_{jp} - \delta_{ij}r_p^2) | I, M_I' \rangle$$

$$= C \langle I, M_I | \tfrac{3}{2}(\mathbf{I}_i\mathbf{I}_j + \mathbf{I}_j\mathbf{I}_i) - \delta_{ij}\mathbf{I}^2 | I, M_I' \rangle \quad (5.98)$$

where C is independent of M_I and M_I' and of the axes i, j. The value of C is obtained by evaluating both sides of (5.98) for any one matrix element. The simplest is for the component $i = j = z$, and the diagonal matrix element for the state $|I, I\rangle$; we have

$$\langle I, I | \sum_p (3z_p^2 - r_p^2) | I, I \rangle = C \langle I, I | 3I_z^2 - I^2 | I, I \rangle. \quad (5.99)$$

The left hand side is defined as the quadrupole moment when the spin axis is aligned to the z-axis, and from (5.91) this is eQ. The right hand side is $C(3I^2 - I(I + 1))$. We can therefore write

$$C = \frac{eQ}{I(2I - 1)}. \quad (5.100)$$

By substituting these expressions into (5.96) we have for the quadrupole interaction Hamiltonian

$$\mathbf{H}_Q = \frac{eQ}{6I(2I - 1)} \sum_{i,j} V_{ij}[\tfrac{3}{2}(\mathbf{I}_i\mathbf{I}_j + \mathbf{I}_j\mathbf{I}_i) - \delta_{ij}\mathbf{I}^2] \quad (5.101)$$

or if we choose our coordinate axes to be the principal axes of the field gradient tensor, this reduces to

$$\mathbf{H}_Q = \frac{eQ}{6I(2I - 1)} \sum_i V_{ii}(3I_i^2 - \mathbf{I}^2). \quad (5.102)$$

As an example we can evaluate the energy of the nucleus in an axially-symmetric field gradient. If we make use of the condition $\eta = 0$, then we can write

$$V_{zz} = -2V_{xx} = -2V_{yy} \quad (5.103)$$

so that the energy of the spin state $|I, M_I\rangle$ is

$$\frac{e^2qQ}{6I(2I - 1)} \langle I, M_I | 3I_z^2 - \tfrac{3}{2}I_x^2 - \tfrac{3}{2}I_y^2 | I, M_I \rangle$$

$$= \frac{e^2qQ}{4I(2I - 1)} [3M_I^2 - I(I + 1)]. \quad (5.104)$$

For a field of axial symmetry the operator $\mathbf{H}_Q$ is diagonal in this set of spin states.

The energy levels are, from (5.104), quadratic in the quantum number M_I, so that a spin 1 or $\frac{3}{2}$ nucleus gives just two energy levels. Transitions between the two lead to a measure of q, or if the quadrupole moment is not known precisely, of the product qQ. The factor e^2qQ/h is generally known as the quadrupole coupling constant. For spins $I > 2$, more than two levels occur. For fields without axial symmetry one can determine also the asymmetry parameter η except for $I = \frac{3}{2}$ nuclei, which, by Kramer's theorem, can only have two energy levels, $|M_I| = \frac{3}{2}, \frac{1}{2}$, in the absence of a magnetic field.

The electric field parameters q and η may also be determined from other branches of spectroscopy, notably pure rotation and Mossbauer spectroscopy. The interest for the theoretical chemist lies in the interpretation of these quantities in terms of molecular wave functions.

The field gradient at a nucleus is made up of a contribution from all the electrons and one from the nuclei. If the electronic wave function is built up from atomic orbitals then the overlap contributions to the density $\varphi_\mu \varphi_\nu$, will give a field gradient at a nucleus A equal to

$$(q_A)_{\mu\nu} = \int \varphi_\mu (3 \cos^2 \theta - 1) r^{-3} \varphi_\nu \, dv \tag{5.105}$$

where r and θ are polar coordinates referred to nucleus A as origin. Integrals of this type are, in general, three-centre, and a little tedious to evaluate. In a ZDO approximation they would be equated to zero unless $\mu = \nu$. An approximate method, originally due to Townes and Dailey[92] and formulated for molecular orbital theory by Gordy[93] and by Cotton and Harris,[94] is frequently used to evaluate these field gradients. This method is to neglect all contributions from the electron density except those arising from atomic orbitals centred on the nucleus in question. The contribution from electron densities on other nuclei is then assumed to be exactly cancelled by the field gradients arising from the other nuclei.

Electrons occupying s orbitals on atom A will have zero field gradient at the nucleus. An electron in a p orbital will give a contribution to q. For a pz orbital this is equal to

$$eq = V_{zz} = \frac{4e}{5} \int R_p^2(r) r^{-3} (r^2 \, dr) \tag{5.106}$$

where R_p is the normalized radial part of the p-orbital wave function. By virtue of the axial symmetry (5.103) this electron will contribute $-eq/2$ to both V_{xx} and V_{yy}. Correspondingly an electron in a px or py orbital will give a contribution $-eq/2$ to V_{zz}. It follows that if there is equal population of the three p orbitals the resulting spherically symmetric electron density makes no contribution to the field gradient.

If there are no orbitals of higher angular momentum than $l = 1$ in the basis then, in the above model, the field gradient depends only on the population of p orbitals and on the radial integrals defined in (5.106). If there are orbitals of higher l, d orbitals for example, then there will be other contributions to equation (5.106) from one-centre densities like $\varphi_s \varphi_d$.

It is usual to assign an empirical value to the integral (5.106) from the quadrupole coupling constant of the free atom, if this is known. Thus in terms of the elements of the molecular bond-order matrix of the atom in question we can write a quadrupole coupling constant

$$C = \frac{e^2 qQ}{h} = C_0[P_{zz}{}^A - \tfrac{1}{2}(P_{xx}{}^A + P_{yy}{}^A)] \qquad (5.107)$$

where C_0 is the value for the atom on the basis of one electron occupying orbital pz.

Values of C_0 for Cl^{35}, Br^{79}, I^{127} have been quoted by Sichel and Whitehead as 109·746, −769·756 and 2292·712 MHz respectively.[95] For N^{14} the ground state is ^4S and so no value of C_0 can be derived from atomic spectral data. A direct calculation of the integral in (5·106) has given $C_0 = -7·4$ MHz but empirical values of -10, -14 MHz have also been used in calculations.[96–98] The problem of deriving C_0 from non-empirical calculations is that no account is usually taken of the so-called Sternheimer polarization of the inner electrons.[99]

The first published calculations of quadrupole coupling constants using the CNDO/2 method were by Davies and Mackrodt.[97] They considered nitrogen couplings and used the one-centre approximation described above and a value $C_0 = -14$ MHz. The molecules considered were pyrrole, pyridine, pyridazine, oxazole and some isoxazoles. The results were generally satisfactory with the exception of pyrrole, for which the field gradient along the axis perpendicular to the molecular plane was calculated to be 2·5 times larger than the experimental estimate. The authors suggest that for this molecule the multicentre contributions to the field gradient may be important. The results were better than those from earlier calculations by the same authors[96] which used only an SCF π method and an estimation of the σ-electron contribution.

Davies and Mackrodt[97] used the bond-order matrix directly obtained from the CNDO/2 eigenfunctions. Betsuyaku[98] used the same method for a calculation on NO_2^- but assumed that the coefficients of these eigenfunctions corresponded to a Löwdin orthogonalized basis set (2.44), and therefore transformed the bond-order matrix to an appropriate non-orthogonal Slater basis. His results, based on the factor $C_0 = -10$ MHz, were as follows:

$$\text{Calc. } C = 5 \cdot 870 \text{ MHz}, \quad \eta = 0 \cdot 50$$
$$\text{Obs. } C = 5 \cdot 791 \text{ MHz}, \quad \eta = 0 \cdot 40$$

with the 2-fold axis of the ion being the z axis. The result is again rather satisfactory, although only η is independent of the value of C_0.

TABLE 5.7 Comparison of calculated and observed quadrupole coupling constants[95] (MHz) based on the parameters $C_0(\text{Cl}) = 109 \cdot 746$, $C_0(\text{Br}) = -769 \cdot 756$, $C_0(\text{N}) = -7 \cdot 4$ MHz. Values in parenthesis are based on Pople–Segal resonance integrals

	Calc.	Obs.
HCl	$-88 \cdot 25$	$-67 \cdot 3$
CH_3Cl	$-78 \cdot 63$	$-74 \cdot 77$
ClF	$-126 \cdot 81$	$-146 \cdot 00$
Cl_2	$-106 \cdot 16$	$-108 \cdot 95$
HBr	$657 \cdot 9$	530
BrF	$952 \cdot 4$	1089
Br_2	$750 \cdot 8$	765
NH_3	$1 \cdot 098 (-4.1)$	$-4 \cdot 084$
N_2	$-0 \cdot 663 (-2.3)$	$-4 \cdot 65$
HCN	$6 \cdot 664 (-1.9)$	$-4 \cdot 58$
FCN	$8 \cdot 823 (-0.6)$	$-2 \cdot 67$

Sichel and Whitehead[95] have given a more extensive analysis of CNDO results and compared them with calculations using the extended Hückel method. The CNDO results were found to be insensitive to the method used to calculate the 2-centre couloumb integrals γ_{AB}. Table 5.7 gives a selection of their results, based on the Mataga–Nishimoto recipe.

The results for the halogens (calculations were also made for some iodides) are in reasonable agreement with experiment: those for nitrogen are poor—contrary to the findings of Davies and Mackrodt[97] and Betsuyaku.[98] However, it is significant that for nitrogen, calculations using the resonance integrals recommended by Pople and Segal, give much better results than those based on the resonance integrals used by Sichel and Whitehead in their papers. The extended Hückel method gave poorer results for the halogens, and, for nitrogen, results comparable with those using the Pople–Segal parameters. Eletr[100] has also compared the CNDO/2 method with extended Hückel calculations (iterated to give one-centre integrals as a function of charge on the atom) for pyrrole, pyridine and pyrazine. He found that the CNDO results (with $C_0 = -8 \cdot 1$ MHz) were not as good as the extended Hückel. He states that his preliminary calculations with the INDO method show better agreement with the experimental results.

Dewar and coworkers[103] have used the MINDO/2 method to calculate the chlorine quadrupole coupling constants of the chlorobenzenes. The MINDO/2 parameters for Cl were calculated by fitting bond lengths and

TABLE 5.8 Comparison of calculated and observed chlorine coupling constants for chlorobenzenes.[103] The calculated values were derived from the expression

$$e^2qQ = AP_{zz}^{Cl} + B$$

where A and B were found by a least squares procedure to be $A = -79 \cdot 636$, $B = 184 \cdot 102$MHz. For non-equivalent chlorine atoms an average was taken

| Compound | P_{zz}^{Cl} | e^2qQ (MHz) | |
		Obs.	Calc.
mono	1·4417	69·24	69·29
para	1·4321	69·56	70·05
meta	1·4273	69·80	70·44
1,2,4	1·4147	72·12	71·44
1,3,5	1·4105	71·70	71·78
ortho	1·4089	71·44	71·70
1,2,3,5	1·3967	73·92	72·88
1,2,3	1·3956	73·10	72·96
1,2,3,4	1·3844	74·68	73·86
1,2,4,5	1·3818	73·60	74·06
penta	1·3665	75·52	75·28
hexa	1·3370	76·88	77·63

heats of formation for some standard molecules, as described in section 3.6. The coupling constant was taken to be proportional to the population of the chlorine $3p$ orbital which is directed along the C—Cl bond, P_{zz}^{Cl}. The results are shown in table 5.8. The general trend with increasing number of chlorine atoms is well reproduced, but the values for different isotopes is in some cases wrong. It is said that the theory does not account well for the asymmetry parameters.

We believe that the results of ZDO calculations on quadrupole coupling constants are rather better than one might expect. The field gradient at a nucleus is a rather sensitive function of the electron distribution in a molecule and *ab initio* calculations have not been particularly successful for this molecular property. The field gradient is sensitive to the orbital exponent of the p orbitals (by virtue of its effect on integrals such as (5.106)), and to other aspects of the basis set.[101] However, it has been found for N_2 that the cancellation of the field gradient contributions from other nuclei and electrons on these nuclei is reasonably complete.[102] The very poor results obtained by Sichel and Whitehead for nitrogen using their bonding parameters appear to us to be significant and to cast doubt on this aspect of their parameterization.

References

1. A. Carrington and A. D. McLachlan, *Introduction to Magnetic Resonance*, Harper and Row, New York, 1967.
2. S. I. Weissman, *J. Chem. Phys.*, **22**, 1135 (1954).
3. H. M. McConnell, *J. Chem. Phys.*, **24**, 632 (1956).
4. H. M. McConnell, *J. Chem. Phys.*, **24**, 764 (1956).
5. R. Bersohn, *J. Chem. Phys.*, **24**, 1066 (1956).
6. S. I. Weissman, *J. Chem. Phys.*, **25**, 890 (1956).
7. A. D. McLachlan, H. H. Dearman and R. Lefebvre, *J. Chem. Phys.*, **33**, 65 (1960).
8. A. Carrington, *Quart. Revs.*, **17**, 67 (1963).
9. G. Hoijtink, *Mol. Phys.*, **1**, 157 (1958).
10. H. M. McConnell and D. B. Chesnut, *J. Chem. Phys.*, **28**, 107 (1958).
11. A. D. McLachlan, *Mol. Phys.*, **3**, 233 (1960).
12. A. Carrington and J. C. P. Smith, *Mol. Phys.*, **9**, 137 (1965).
13. H. R. Falle and G. R. Luckhurst, *Mol. Phys.*, **11**, 299 (1966).
14. I. Bernal, P. H. Rieger and G. K. Fraenkel, *J. Chem. Phys.*, **37**, 1489 (1962).
15. A. Hinchliffe, *Theoret. Chim. Acta*, **5**, 208 (1966).
16. L. C. Snyder and A. T. Amos, *J. Chem. Phys.*, **42**, 3670 (1965).
17. J. A. Pople and R. K. Nesbet, *J. Chem. Phys.*, **22**, 571 (1954).
18. A. Brickstock and J. A. Pople, *Trans. Faraday Soc.*, **50**, 901 (1954).
19. H. G. Benson and A. Hudson, *Mol. Phys.*, **20**, 185 (1971).
20. P. J. Black and C. A. McDowell, *Mol. Phys.*, **12**, 233 (1967).
21. M. A. Ali and A. Hinchliffe, *Trans. Faraday Soc.*, **62**, 3273 (1966).
22. R. W. Fessenden and S. Ogawa, *J. Amer. Chem. Soc.*, **86**, 3591 (1964).
23. R. G. Lawler, J. R. Bolton, G. K. Fraenkel and T. H. Brown, *J. Amer. Chem. Soc.*, **86**, 520 (1964).
24. W. D. Hobey, *Mol. Phys.*, **7**, 325 (1964).
25. M. Karplus, R. G. Lawler and G. K. Fraenkel, *J. Amer. Chem. Soc.*, **87**, 5260 (1965).
26. J. E. Bloor, B. R. Gilson and P. N. Daykin, *J. Phys. Chem.*, **70**, 1457 (1966).
27. J. R. Bolton, *J. Chem. Phys.*, **43**, 309 (1965).
28. J. P. Malrieu, *J. Chem. Phys.*, **46**, 1654 (1967).
29. K. D. Sales, 'Theory of Isotropic Hyperfine Splitting Constants for Organic Free Radicals', in *Advances in Free Radical Chemistry*, Vol. 3, Logos Press, 1969, p. 139.
30. E. de Boer and J. P. Colpa, *J. Phys. Chem.*, **71**, 21 (1967).
31. M. Karplus and G. K. Fraenkel, *J. Chem. Phys.*, **35**, 1312 (1961).
32. J. C. M. Henning, *J. Chem. Phys.*, **44**, 2139 (1966).
33. A. Hinchliffe and J. N. Murrell, *Mol. Phys.*, **14**, 147 (1968).

34. M. T. Melchior, *J. Chem. Phys.*, **50**, 511 (1969).
35. M. Broze and Z. Luz, *J. Chem. Phys.*, **51**, 738 (1969).
36. M. D. Newton and W. E. Palke, *J. Chem. Phys.*, **45**, 2329 (1966).
37. J. A. Pople, D. L. Beveridge and P. A. Dobosh, *J. Amer. Chem. Soc.*, **90**, 4201 (1968).
38. C. Thomson, *Theoret. Chim. Acta*, **17**, 320 (1970).
39. W. C. Danen and T. T. Kensler, *J. Amer. Chem. Soc.*, **92**, 5235 (1970).
40. P. Bakuzis, J. K. Kochi and P. J. Krusic, *J. Amer. Chem. Soc.*, **92**, 1434 (1970).
41. P. H. Kasai, P. A. Clark, and E. B. Whipple, *J. Amer. Chem. Soc.*, **92**, 2640 (1970).
42. P. J. Krusic and T. A. Rettig, *J. Amer. Chem. Soc.*, **92**, 722 (1970).
43. J. A. Pople, W. G. Schneider and H. J. Bernstein, *High Resolution Nuclear Magnetic Resonance*, McGraw-Hill, 1959.
44. J. W. Emsley, J. Feeney and L. H. Sutcliffe, *High Resolution N.M.R. Spectroscopy*, Pergamon, 1965.
45. E. O. Bishop, *Ann. Rev. N.M.R.* **1**, 91 (1967).
46. N. F. Ramsey, *Phys. Rev.*, **78**, 699 (1950).
47. N. F. Ramsey and E. M. Purcell, *Phys. Rev.*, **85**, 143 (1952).
48. N. F. Ramsey, *Phys. Rev.*, **91**, 303 (1953).
49. M. Barfield and D. M. Grant, *Advances in Magnetic Resonance*, Vol. 1, Academic, 1965, p. 149.
50. J. N. Murrell, *Progr. N.M.R.*, **6**, 1, 1970.
51. T. P. Das and R. Bersohn, *Phys. Rev.*, **104**, 849 (1956).
52. A. Saika and C. P. Slichter, *J. Chem. Phys.*, **22**, 26 (1954).
53. J. A. Pople, *Proc. Roy. Soc.*, **A239**, 541 (1957); *J. Chem. Phys.*, **37**, 53 (1962); *Disc. Faraday Soc.*, **34**, 7 (1962).
54. W. E. Lamb, *Phys. Rev.*, **60**, 817 (1941).
55. C. W. Kern and W. N. Lipscomb, *J. Chem. Phys.*, **37**, 260 (1962).
56. A. J. Sadlej, *Mol. Phys.*, **19**, 749 (1970).
57. J. A. Pople, *J. Chem. Phys.*, **24**, 1111 (1956).
58. J. S. Waugh and R. W. Fessenden, *J. Amer. Chem. Soc.*, **79**, 846 (1957).
59. C. E. Johnson and F. A. Bovey, *J. Chem. Phys.*, **29**, 1012 (1958).
60. J. A. Pople, *Mol. Phys.*, **1**, 175 (1958).
61. R. McWeeny, *Mol. Phys.*, **1**, 311 (1958).
62. F. London, *J. Phys. Radium*, **8**, 397 (1937).
63. G. G. Hall and A. Hardisson, *Proc. Roy. Soc.*, **A268**, 328 (1962).
64. G. G. Hall, A. Hardisson and L. M. Jackman, *Tetrahedron*, S2, 101 (1963).
65. G. G. Hall, A. Hardisson and L. M. Jackman, *Disc. Faraday Soc.*, **34**, 15 (1962).
66. W. Adam, A. Grimison and G. Rodriguez, *J. Chem. Phys.*, **50**, 645 (1969).
67. J. B. Lambert and J. D. Roberts, *J. Amer. Chem. Soc.*, **87**, 4087 (1965).
68. A. Velenik and R. M. Lynden-Bell, *Mol. Phys.*, **19**, 371 (1970).
69. M. Weeks, Private communication.
70. D. W. Davies, *Mol. Phys.*, **13**, 465 (1967).
71. R. Ditchfield, D. P. Miller and J. A. Pople, *J. Chem. Phys.*, **53**, 613 (1970).
72. J. A. Pople and D. P. Santry, *Mol. Phys.*, **8**, 1 (1964).
73. J. A. Pople and D. P. Santry, *Mol. Phys.*, **7**, 269 (1963).
74. R. Hoffmann, *J. Chem. Phys.*, **39**, 1397 (1963).

75. R. Ditchfield and J. N. Murrell, *Mol. Phys.*, **14**, 481 (1968).
76. R. Ditchfield, *Mol. Phys.*, **17**, 33 (1969).
77. P. Loéve and L. Salem, *J. Chem. Phys.*, **43**, 3402 (1965).
78. E. A. G. Armour and A. J. Stone, *Proc. Roy. Soc.*, **A302**, 25 (1967).
79. J. A. Pople, J. W. McIver and N. S. Ostlund, *Chem. Phys. Letters*, **1**, 465 (1967); *J. Chem. Phys.*, **49**, 2960, 2965 (1968).
80. H. D. Cohen and C. C. J. Roothaan, *J. Chem. Phys.*, **43**, 534 (1965).
81. R. Ditchfield, N. S. Ostlund, J. N. Murrell and M. A. Turpin, *Mol. Phys.*, **18**, 433 (1970).
82. A. C. Blizzard and D. P. Santry, *Chem. Comm.*, 1085 (1970).
83. J. N. Murrell, P. E. Stevenson and G. T. Jones, *Mol. Phys.*, **12**, 265 (1967).
84. E. A. G. Armour, *J. Chem. Phys.*, **49**, 5445 (1968).
85. W. H. de Jeu, *Mol. Phys.*, **20**, 573 (1971).
86. Y. Kato and A. Saika, *J. Chem. Phys.*, **46**, 1975 (1967).
87. J. N. Murrell, M. A. Turpin and R. Ditchfield, *Mol. Phys.*, **18**, 271 (1970).
88. A. Hinchliffe and D. B. Cook, *Theoret. Chim. Acta*, **17**, 91 (1970).
89. A. D. Buckingham, *Quart. Rev.*, **13**, 183 (1959).
90. T. P. Das and E. L. Hahn, *Nuclear Quadrupole Resonance Spectroscopy*, Academic, New York, 1958.
91. A. Abragam, *The principles of nuclear magnetism*, Oxford, 1961.
92. C. H. Townes and B. P. Dailey, *J. Chem. Phys.*, **17**, 782 (1949).
93. W. Gordy, *Disc. Faraday Soc.*, **19**, 14 (1955).
94. F. A. Cotton and C. B. Harris, *Proc. Nat. Acad. Sci.*, **56**, 12 (1966).
95. J. M. Sichel and M. A. Whitehead, *Theoret. Chim. Acta*, **11**, 263 (1968).
96. D. W. Davies and W. C. Mackrodt, *Chem. Comm.*, **1967**, 345.
97. D. W. Davies and W. C. Mackrodt, *Chem. Comm.*, **1967**, 1226.
98. H. Betsuyaku, *J. Chem. Phys.*, **50**, 3117, 3118 (1969).
99. R. M. Sternheimer, *Phys. Rev.*, **84**, 244 (1951); **86**, 316 (1952); **95**, 736 (1954).
100. S. Eletr, *Mol. Phys.*, **18**, 119 (1970).
101. S. Eletr, T. K. Ha, C. T. O'Konski, *J. Chem. Phys.*, **51**, 1430 (1969).
102. J. W. Richardson, *Rev. Mod. Phys.*, **32**, 461 (1960).
103. M. J. S. Dewar, D. H. Lo, D. B. Patterson, N. Trinajstić and G. E. Peterson, *Chem. Comm.*, **1971**, 238.

Chapter 6

Future developments

In speculating on the future direction of semi-empirical SCF–MO theory it is probably important to distinguish between two aims which we have mentioned in earlier chapters. These are to reproduce the results of non-empirical SCF calculations and to reproduce experimental data. These aims overlap to some extent because there are some results of non-empirical SCF calculations which one expects to be generally in good agreement with experiment, for example, equilibrium geometries and dipole moments. There are, however, other aspects of such calculations, notably those associated with the correlation-energy problem, where the SCF results may be in poor agreement with experiment. We think that arguments can be given in favour of both aims and we therefore discuss separately the likely steps to be taken by both protagonists.

6.1 The approach to non-empirical SCF calculations

At the present time it is a relatively routine matter to do a non-empirical SCF calculation with a minimum or slightly extended basis of Slater orbitals on a small polyatomic molecule such as formaldehyde (H_2CO). The calculations can either be done directly with Slater orbitals or one can use equivalent small gaussian orbital expansions for each Slater orbital. The facility to do such calculations relies on access to a computer of at least 32k store. Semi-empirical calculations on such small molecules at the present time would have little significance unless they were being used to test a method which is aimed at larger molecules or unless they were parameterized to give better results than the non-empirical calculations.

It must be remembered than when we adopt the LCAO approximation for a molecular orbital

$$\psi = \sum_{\mu=1}^{n} c_\mu \varphi_\mu \tag{6.1}$$

we do so in the hope that it is a reasonably good method of expanding a very complicated function in three dimensional space. The length (n) of the expansion required to obtain a specified accuracy to the energy of ψ depends on the nature of the φ_μ. Thus more gaussian orbitals are required than Slater orbitals. The compromise in any LCAO method is between the length of the expansion set and the difficulty of evaluating the F-matrix elements for the basis functions. At one extreme there are single-centre expansions for which the integrals are very easy to evaluate but the rate of convergence is very poor.

TABLE 6.1   Mean absolute errors for bond lengths and bond angles

	INDO	STO-3G	LEMAO-4G
Bond length Å	0·035	0·011	0·028
Bond angles (°)	1·4	2·0	3·1

STO-NG is a method which replaces each Slater orbital by N gaussians fitted by a least-squares technique.
LEMAO-NG uses N gaussians per atomic orbital chosen by minimizing the energy of the atom.

The evidence at present available would suggest that SCF calculations using a minimum basis of Slater orbitals give equilibrium geometries which in most cases are in reasonable agreement with experiment. There is less support for calculations which use a *minimum* basis of gaussian orbitals. In table 6.1 we give mean absolute errors quoted by Pople[1] for some small molecules (H_2O, NH_3, CH_4, HCN etc.) using the INDO method and two different Gaussian expansion *ab initio* methods. Likewise the electron densities, in so far as they are tested against measured dipole moments are probably reasonable.

The heat of atomization from such a calculation is, on the other hand, generally in poor agreement with experiment because of the failure to account for the correlation energy. However, if we take advantage of the fact that correlation energy appears to be roughly the same value for each electron-pair bond ($\sim$1 eV), then we can calculate reaction energies as the difference in SCF energies, providing the number of electron-pair bonds is conserved in the reaction.[2,3] For example, the reaction

$$H_2 + Li_2 \rightarrow 2LiH$$

has a standard heat of reaction of $16·6 \pm 1·2$ kcal/mole and an SCF calculation gives 16·5 kcal/mole.

Having established that non-empirical SCF calculations have experimental significance we can ask how they might be approached for large molecules.

Two directions appear worthy of study. The first is to take explicit account of all electrons but to make such approximations to the integrals as will reduce substantially the time necessary for the calculation. The zero-overlap approximations described in this book have this as one of their aims but the approximations are so severe, that unless they are compensated by the use of empirical parameters at other points in the calculation, the results will bear little resemblance to those of a non-empirical calculation. The second approach is to neglect inner-shell electrons, and we will return to this topic later.

Most integral approximations which are both simple and have a reasonable chance of success are based on some variant of the Mulliken approximation. Under this heading one usually includes all approximations which involve the replacement of an overlap density $\varphi_\mu \varphi_\nu$ by a sum of the one electron densities φ_μ^2 and φ_ν^2 suitably scaled so that the total integrated density is the same. Thus we can replace $\varphi_\mu \varphi_\nu$ by

$$\theta_{\mu\nu} = S_{\mu\nu}[a^2 \varphi_\mu^2 + b^2 \varphi_\nu^2] \qquad (6.2)$$

noting that if

$$\int \varphi_\mu \varphi_\nu \, dv = \int \theta_{\mu\nu} \, dv \qquad (6.3)$$

then $a^2 + b^2 = 1$. This has most frequently been used in its simplest form with $a^2 = b^2 = \frac{1}{2}$, but there are other recipes such as that of Löwdin[4] who recommends that a and b be determined by equating the first moments of $\varphi_\mu \varphi_\nu$ and $\theta_{\mu\nu}$ in the direction of the M—N bond

$$\int \varphi_\mu \varphi_\nu \cdot x \, dv = S_{\mu\nu} \int (a^2 \varphi_\mu^2 + b^2 \varphi_\nu^2) x \, dv, \qquad (6.4)$$

the origin of x being the centre of charge of $\varphi_\mu \varphi_\nu$.

The Mulliken approximation for resonance integrals (2.29), or more precisely the Wolfsberg–Helmholz approximation (2.30) are of this family. However, from the point of view of saving time in a calculation the more important approximation is the following for electron-repulsion integrals:

$$(\mu\nu \mid \rho\sigma) = \tfrac{1}{4} S_{\mu\nu} S_{\rho\sigma}[(\mu^2 \mid \rho^2) + (\mu^2 \mid \sigma^2) + (\nu^2 \mid \rho^2) + (\nu^2 \mid \sigma^2)] \quad (6.5)$$

In investigations of the Mulliken approximation it has been generally agreed that it is poor for the kinetic energy integrals.[5] Thus

$$\int \varphi_\mu \, \nabla^2 \varphi_\nu \, dv \qquad (6.6)$$

is very far from being equal to

$$\tfrac{1}{2}S_{\mu\nu}\left[\int \varphi_\mu \, \boldsymbol{\nabla}^2\varphi_\mu \, dv + \int \varphi_\nu \, \boldsymbol{\nabla}^2\varphi_\nu \, dv\right] \tag{6.7}$$

However, as these integrals are no more difficult to evaluate than the overlap integrals themselves there would be little computational significance in such a replacement. Also, if the Mulliken approximation is used for different orbitals on the same centre the resulting integrals will be zero because these orbitals are orthogonal. However, one-centre exchange integrals like $(sp \mid sp)$ may be quite large.

Finally, it should be emphasized that the Mulliken approximation does not lead to results which are invariant to an orthogonal transformation of the atomic orbital basis. Quite different results may be obtained when it is applied to integrals involving hybrid orbitals than when it is applied to integrals involving atomic orbitals.

It is invidious to pick out a few of the many studies based on the Mulliken approximation but we take the risk. In π-electron theory there is the work of Hummel and Ruedenberg on the spectra of aromatic hydrocarbons.[6] For small molecules there is the attempt to develop a noniterative MO theory by Lipscomb and coworkers in which the Mulliken approximation is used.[7] Finally a review article by Nicholson studies the Mulliken approximation amongst many other approaches to empirical SCF theory.[8]

Table 6.2 shows some calculations carried out by Nicholson with the object of examining the errors arising through the Mulliken approximation. Calculations I are based on the Mulliken approximation applied strictly to all electron-repulsion integrals, and to the one-electron integrals except for kinetic energies. Calculations II add some one-centre two-electron exchange integrals which are zero in calculation I, and calculation III adds to calculation I the correct one-centre off-diagonal F-matrix elements such as F_{sp} which are also zero in the Mulliken approximation.

By comparing the results of the different calculations in table 6.2 one is comforted to find that at least the order of the molecular orbitals is the same in all cases. This is not however true of the vacant orbitals; that is, the eigenfunctions of the F matrices which are not occupied by electrons in the ground states of the molecules are not in the same order for all calculations.

The major difference between the three Mulliken-based calculations and the *ab initio* results lies in the electron distribution. The one-centre exchange integrals clearly have a big influence on this but none of the approximate calculations gets close to reproducing the *ab initio* results.

TABLE 6.2 Calculations to test the Mulliken approximation. Energies (eV) are given for the top 3 or 4 bonding orbitals and the symmetries of these orbitals are indicated. The Mulliken population numbers defined by

$$q_\mu = \sum_\nu P_{\mu\nu} S_{\mu\nu},$$

are quoted for the hydrogen $1s$ orbitals and the $2s$ orbitals Details of the calculations are given in the text

	Accurate[14]	I[8]	II[8]	III[8]	IV[9]
NH_3					
$-E(a_1)$	29·96	31·45	33·52	31·48	—
$-E(e)$	15·84	16·22	16·22	16·14	—
$-E(a_1)$	9·96	8·16	9·74	6·91	—
q_h	0·84	1·06	0·91	1·05	—
q_s	1·59	1·10	1·28	1·09	—
C_2H_2					
$-E(\sigma_g)$	27·35	—	—	—	28·95
$-E(\sigma_u)$	20·43	22·28	24·05	20·98	27·24
$-E(\sigma_g)$	17·85	14·72	15·43	15·92	23·46
$-E(\pi_u)$	11·05	9·17	9·20	10·29	10·34
q_h	0·81	0·89	0·75	1·02	0·90
q_s	1·10	1·12	1·22	1·06	0·92
C_2H_4					
$-E(b_u)$	17·52	17·96	18·07	18·31	—
$-E(a_g)$	15·29	12·76	13·01	13·63	—
$-E(b_{2g})$	13·77	11·46	11·40	11·73	—
$-E(b_{2u})$	10·09	9·31	10·20	9·85	—
q_h	0·86	1·03	0·90	1·06	—
q_s	1·19	1·06	1·19	1·04	—

Nicholson gives a thorough analysis of the failure of the approximate calculations which is not easy to summarize. The most important feature appears to be an under-valuation of the two-centre exchange integrals $(\varphi_A \varphi_B \mid \varphi_A \varphi_B)$ which appear in the diagonal elements of the F matrix.

If one is simply seeking to save significant time in the evaluation of integrals required to build up the F-matrix elements one could reasonably limit approximations to the three and four-centre integrals alone, since these are in the first place the most difficult to calculate and secondly they are amongst the smallest integrals of a calculation. Unfortunately there are many of them, so that unless there is some cancellation of errors one could still be in serious error overall. Table 6.2 (column IV) includes some calculations by Burnelle[9] on acetylene with such approximations. In addition some of the very small three and four-centre integrals were taken as zero. Unfortunately the energies cannot be directly compared with the others in the table as the atomic orbital exponents for the hydrogen $1s$

and the carbon $2s$ and $2p$ orbitals are different ($1 \cdot 0$ instead of $1 \cdot 2$ and $1 \cdot 59$ instead of $1 \cdot 625$). It is this different choice of exponents which presumably has such a large influence on the energies of the σ_u and σ_g orbitals. Burnelle showed that with his method of calculation the geometry of the first excited state was not correctly given, and concluded that this was possibly due to the approximations made in the integrals. This is an important question: do *ab initio* calculations with the Mulliken approximation give geometries that are at least as good as by the ZDO methods we have described in chapter 3?

We turn now to the other time-consuming aspect of calculations for heavy elements and that is on the treatment of inner-shell electrons. If it were possible to do *ab initio* SCF calculations on valence-shell electrons alone then we could treat transition metal complexes as readily as we can now treat hydrocarbons. Although the weight of chemical evidence suggests that a detailed consideration of inner shells should not be necessary to understand most chemical properties, it appears difficult to remove them from the atomic orbital basis without having a drastic effect on the resulting valence molecular orbitals.

The first problem we encounter when we ignore inner-shell orbitals is that the valence atomic orbital basis must now approximate to SCF atomic orbitals by having radial nodes. If inner shells are included in the basis then we can ignore this factor because radial nodes are introduced through the overlap integrals between valence and inner orbitals of the same atom.

For example, if we have a basis of nodeless Slater orbitals, $1s$ and $2s$, then these are not orthogonal. We can however find a noded $2s$ orbital which is orthogonal to the $1s$ by taking a linear combination of $1s$ and $2s$ as follows

$$\varphi_{2s}' = \varphi_{2s} - S_{1s2s}\varphi_{1s}. \tag{6.8}$$

This is the so-called Schmidt orthogonalization. We can now confirm that

$$S_{1s2s'} = 0 \tag{6.9}$$

and it is found that ϕ_{2s}' is a good approximation to a spectroscopic $2s$ atomic orbital of the atom.

If in a molecular orbital we have a linear combination of $1s$ and $2s'$, then this can clearly be replaced by a linear combination of $1s$ and $2s$. From (6.8) we have

$$a\varphi_{1s} + b\varphi_{2s}' = (a - bS_{1s2s})\varphi_{1s} + b\varphi_{2s}. \tag{6.10}$$

Thus it makes no difference whether our atomic orbital basis consists of $1s$ and $2s$ or $1s$ and $2s'$. However, if we ignore the $1s$ orbital in our basis then there is a difference between the basis with $2s$ and that with $2s'$.

The second problem that one faces if inner shells are ignored is that there is not a large enough difference between the energies of atomic orbitals with the same principal quantum number but different l quantum numbers (e.g. $2s$ and $2p$). One might say that the main difference in energy between a $2s$ and $2p$ orbital arises from the fact that a $2s$ orbital has a greater penetration of the $1s$ core and therefore experiences a greater effective nuclear charge. In other words it is a poor approximation to treat inner shells simply by replacing the nuclear potential by the coulomb potential of an effective nuclear charge,

$$-Z_{eff}(e^2/r). \tag{6.11}$$

Nicholson has established that if for the elements B to F one takes $Z_{eff} = Z - 2$ it is necessary to add an additional stabilization energy of $-0\cdot063Z_{eff}$ (a.u.) for $2s$ orbitals and $-0\cdot007Z_{eff}$ (a.u.) for $2p$ orbitals. No doubt comparable correction factors could be established for heavier atoms.

Penetration effects for orbitals on other atoms can usually be neglected. Thus a two-centre nuclear-attraction integral is given quite accurately by the integral

$$\int \varphi_a(-Z_{eff}^A/r_A)\varphi_b \, dv. \tag{6.12}$$

We would say that the prospects are good for developing an approximate non-empirical SCF–MO technique whose results would be as useful as those of say the INDO method but which required not much more computer time. However, the efforts that have been spent so far in this direction show that it is not an easy task, and that it will be very difficult to get a simple technique which is very close to accurate *ab initio* calculations both in energies and electron densities.

6.2   The development of empirical SCF theories

The integrals that determine the SCF orbitals can be divided into three categories, one-centre, two-centre-coulomb and those that are overlap dependent. It can be assumed that most empirical theories will take the one-centre terms from atomic spectral data. The two-electron integrals are either calculated, or are calculated with an empirical correction imposed so that they go to the correct one-centre limits. In any calculation it is necessary to decide whether the orbital basis consists of atomic orbitals or orthogonalized atomic orbitals defined by (2.44).

In all the ZDO theories described in this book the only overlap-dependent integral which is included in the F-matrix elements is the resonance integral $\beta_{\mu\nu}$. In almost all the models we have described, $\beta_{\mu\nu}$ has been chosen solely on the basis of optimizing the agreement with either accurate *ab initio* calculations or with experimental data. The problem that one encounters with a general molecule is that there may be a large number of these parameters in the calculation.

The invariance criteria used in the CNDO and INDO methods reduce the number of parameters by taking $\beta_{\mu\nu} = (\beta_M{}^0 + \beta_N{}^0)S_{\mu\nu}$ where $\beta_N{}^0$ depends only on the nature of the atom N. However, we have seen that when these theories were extended to second-row atoms this restriction had to be relaxed to get satisfactory results.

One can anticipate that in ZDO calculations on transition metal complexes (to date there have been few such calculations)[10,11] one will need different β^0 for resonance integrals involving s and p and d orbitals of the metal. The Wolfsberg–Herlmholtz expression (2.30) provides a recipe for relating these parameters but in the independent-electron calculations, for which it was derived, it was not considered very successful. It remains to be seen if it is more successful within the SCF methods.

Formally $\beta_{\mu\nu}$ is well defined in the ZDO theories by expression (2.10). However, if one proceeds to calculate it from this, the integrals being as easy as overlap integrals, then the values obtained are much larger in magnitude than the empirical values found to be optimum in such calculations. The difficulty lies in the fact that $\beta_{\mu\nu}$ is an overlap-dependent integral and in a strict ZDO scheme should be neglected. If it is going to be calculated then one should consider whether there are not other integrals of comparable magnitude which should be taken into account. The following analysis shows that there are.

Let us start with the complete expression for the off-diagonal elements of the F matrix, μ and ν being on different atoms (1.15)

$$F_{\mu\nu} = H_{\mu\nu}{}^c + \sum_\rho \sum_\sigma P_{\rho\sigma}[(\mu\nu \mid \rho\sigma) - \tfrac{1}{2}(\mu\rho \mid \nu\sigma)]. \qquad (6.13)$$

We now collect from this all terms in which the overlap density $\varphi_\mu\varphi_\nu$ is involved but no other overlap density, they are as follows

$$\langle\mu| -\tfrac{1}{2}\nabla^2 + V_M + V_N + \sum_A{}'' V_A |\nu\rangle$$
$$+ \sum_\rho P_{\rho\rho}(\mu\nu \mid \rho^2) - \tfrac{1}{2}P_{\mu\mu}(\mu^2 \mid \mu\nu) - \tfrac{1}{2}P_{\nu\nu}(\mu\nu \mid \nu^2). \qquad (6.14)$$

One can see here an important balance between nuclear attraction and electron repulsion integrals. For example if a new atom L is added to the system then the terms

$$\langle \mu | V_L | \nu \rangle \quad \text{and} \quad \sum_{\rho(L)} P_{\rho\rho}(\mu\nu \,|\, \rho^2) \qquad (6.15)$$

will cancel if L is far from the centres of μ and ν and if L is neutral so that

$$\sum_{\rho(L)} P_{\rho\rho} = Z_L. \qquad (6.16)$$

Thus one cannot attempt a non-empirical calculation of β simply from the nuclear attraction terms unless integrals like $(\mu\nu \,|\, \rho^2)$ are also taken into account.

It is possible to collect the integrals in (6.14) into one parameter if a relationship analagous to (2.18) is introduced as follows (ρ on A)

$$V_{A,\mu\nu} = -F(R)Z_A(\mu\nu \,|\, \rho^2). \qquad (6.17)$$

For the ZDO theories we showed that penetration effects could be ignored for the coulomb terms, that is $F(R) = 1$ if $\mu = \nu$. However, this is unlikely to be true for the overlap-dependent integrals because nucleus A may be the centre for orbital μ or ν. As an approximation we shall take $F(R) = 1$ if μ or ν is not on centre A, and $F(R) = k^{-1}$ (a constant) if one or the other is on centre A. The terms in (6.14) therefore become

$$\langle \mu | -\tfrac{1}{2}\nabla^2 + V_M\left(1 - kZ_M^{-1}\left(\sum_{\rho(M)} P_{\rho\rho} - \tfrac{1}{2}P_{\mu\mu}\right)\right)$$
$$+ V_N\left(1 - kZ_N^{-1}\left(\sum_{\rho(N)} P_{\rho\rho} - \tfrac{1}{2}P_{\nu\nu}\right)\right) + \sum_A'' V_A\left(1 - Z_A^{-1}\sum_{\rho(A)} P_{\rho\rho}\right) | \nu \rangle$$
$$(6.18)$$

or, introducing the net atom charges (3.11),

$$\langle \mu | -\tfrac{1}{2}\nabla^2 + V_M(1 - kZ_M^{-1}(P_{MM} - \tfrac{1}{2}P_{\mu\mu}))$$
$$+ V_N(1 - kZ_N^{-1}(P_{NN} - \tfrac{1}{2}P_{\nu\nu})) + \sum_A'' V_A(1 - Z_A^{-1}P_{AA}) | \nu \rangle. \qquad (6.19)$$

If one neglects the potential from distant atoms ($A \neq M, N$) then one has an expression which could be taken to define the parameter $\beta_{\mu\nu}$, namely

$$\beta_{\mu\nu} = \langle \mu | -\tfrac{1}{2}\nabla^2 + V_M[1 - kZ_M^{-1}(P_{MM} - \tfrac{1}{2}P_{\mu\mu})]$$
$$+ V_N[1 - kZ_N^{-1}(P_{NN} - \tfrac{1}{2}P_{\nu\nu})] | \nu \rangle. \qquad (6.20)$$

This definition would make $\beta_{\mu\nu}$ a function of atoms M and N alone. It would, however, differ from the usual ZDO definition in depending on the electron densities on the two atoms.

To evaluate expression (6.20) one would take $V_M = -Z_M e^2/r$ and calculate the kinetic energy and two-centre nuclear-attraction integrals, which is relatively easy. In preliminary studies with such an expression[12] within the INDO framework we found that if $k = 1$ no convergent SCF solution could be obtained. This appears to be due to an over-estimation of electron repulsion so that electrons tend to get as far from one another as possible, which is not a normal bonding situation. For k less than about 0·8, convergent solutions were obtained, but the values of β were in general appreciably larger than the empirical parameters found to be optimum in the INDO method. Thus the SCF eigenvalues were spread rather more than in the normal INDO calculations.

We have made these comments about $\beta_{\mu\nu}$ to emphasize the problem of defining it non-empirically, rather than to suggest that expression (6.20) is a completely satisfactory definition. Moreover, if one is going to calculate overlap terms then one has to consider their appearance in the diagonal elements of the $\mathbf{F}$ matrix, and in the $\mathbf{S}$ matrix itself.

Nicholson,[8] in his analysis of the ZDO methods, attributes their success to a cancellation of electron-repulsion and electron-nuclear attraction terms in the diagonal elements of the $\mathbf{F}$ matrix, and to the use of empirical $\beta_{\mu\nu}$ in the off-diagonal elements. One of the features he criticizes is the use of spectroscopic data to define the atomic energies, but with overlap integrals being calculated from Slater rather than Hartree–Fock orbitals. Hartree–Fock orbitals are, in general, more diffuse than Slater orbitals and their overlap integrals, particularly at large internuclear separations, can be much greater. He cites the example of HF for which minimum basis SCF calculations give the charge on the hydrogen as $+0·36$ if Hartree–Fock orbitals are used but only $0·10$ if Slater orbitals are used. However, this may be an exceptional case for there is no widespread opinion that in *ab initio* calculations Slater orbitals with well chosen exponents give much poorer results than do Hartree–Fock orbitals.† Orbital exponents have for several cases been optimized for particular molecules (for example reference 13), and compromise exponents for a range of molecules have been proposed.[1] These are listed in table 6.3. We doubt that Hartree–Fock orbitals, which are generally expressed as a sum of Slater orbitals, will be widely adopted in semi-empirical calculations.

A semi-empirical theory based on the Mulliken approximation has been proposed by Nicholson.[8] He uses a rotational average of certain

---

† Roby and Sinanôğlu[15] have considered the question of scaling integrals over Slater orbitals to give integrals over Hartree–Fock orbitals. NDDO calculations on CO showed an encouraging approach to the Hartree–Fock limit.

TABLE 6.3 Compromise exponents (ζ) for Slater orbitals in molecular calculations, compared with Slater's atomic values[1]

	Compromise molecular exponents	Slater atomic exponents
H	1·24	1·00
C	1·72	1·625
N	1·95	1·95
O	2·25	2·275
F	2·55	2·60

integrals to ensure invariance to rotations. The non-spectroscopic empirical parameters in the theory arise in the off-diagonal $\mathbf{F}$-matrix elements. In the first place a parameter k^V is defined by

$$V_{\mathrm{A},\mu\nu} = \tfrac{1}{2}k^V S_{\mu\nu}[V_{\mathrm{A},\mu\mu} + V_{\mathrm{A},\nu\nu}]. \tag{6.21}$$

Secondly, the coulomb matrix elements

$$J_{\mu\nu} = \sum_{\rho}\sum_{\sigma} P_{\rho\sigma}(\mu\nu \mid \rho\sigma), \tag{6.22}$$

are, after applying the Mulliken approximation, (6.5), given by

$$J_{\mu\nu} = \tfrac{1}{2}S_{\mu\nu}\sum_{\rho}{}' N_{\rho}[(\mu\mu \mid \rho\rho) + (\nu\nu \mid \rho\rho)], \tag{6.23}$$

where N_{ρ} is the Mulliken orbital population,

$$N_{\rho} = \sum_{\sigma} P_{\rho\sigma}S_{\rho\sigma}. \tag{6.24}$$

Nicholson introduces a second parameter k^J to multiply those terms in (6.23) which involve orbitals ρ on the same atoms as μ or ν.

The parameters k^V and k^J depend on the type of orbital (μ and ν) but not on the atom involved. Table 6.4 shows the parameters deduced by Nicholson.

Further details of the Nicholson procedure are not given here as the

TABLE 6.4 Mulliken correction factors for hydrogen and the first row elements

μ–ν	1s–1s	1s–2s	1s–2pσ	2s–2s	2s–2pσ	2pσ–2pσ	2pπ–2pπ
k^V	0·75	0·84	0·84	0·85	0·91	1·15	0·89
k^J	0·65	0·94	0·83	0·98	1·00	1·10	0·97

technique has not been tested on a wide range of systems. The first results show satisfactory agreement with the *ab initio* calculations and with experimental results. We have outlined the method primarily to emphasize that empirical methods superimposed on approximations of the Mulliken type are likely to develop in the future in parallel with the more sophisticated ZDO methods like INDO and NDDO.[15]

6.3 Final comments

Because this book was written primarily with the experimental chemist in mind we would like to direct our final remarks to him.

Quantum chemistry has two aims. The first is to provide an understanding of chemical phenomena and a conceptual framework or language to aid communication. The second is to provide theories which may be applied to a wide variety of molecules, without using undue computing time, and whose predictions are reliable and chemically significant. In other words, it must give relative energies, for example, within a few kcal/mole of experimental measurements.

In general, the simpler the theory the more likely it is to provide concepts that chemists will take as part of their language. Concepts like 'hybridization,' and 'molecular orbital,' come from very simple theories. Unfortunately very simple theories are not generally reliable in their predictions—which is not to say that a more complicated theory is necessarily more reliable.

At the present time the empirical SCF theories that we have described come closest to the aim of producing widely available accurate results. Because they are mathematically simple there is a natural tendency to translate their results into explanations. We believe that such translations must be made with great care. A carefully parameterized theory may give the right results for the wrong reasons. Such a theory can be considered as a useful tool of chemistry, but is not necessarily a theory that will increase our understanding of chemistry.

The next few years will undoubtedly confirm the success already achieved by some of the semi-empirical SCF theories. We are entering an era in which in some situations it will be more reliable, and far cheaper, to do an experiment on a computer than on a laboratory bench.

References

1. J. A. Pople, *Accnts. Chem. Res.*, **3**, 217 (1970).
2. L. C. Snyder, *J. Chem. Phys.*, **46**, 3602 (1967); L. C. Snyder and H. Basch, *J. Amer. Chem. Soc.*, **91**, 2189 (1969).
3. R. Ditchfield, W. J. Hehre, J. A. Pople and L. Radom, *Chem. Phys. Letters*, **5**, 13 (1970).
4. P. O. Löwdin, *J. Chem. Phys.*, **21**, 374 (1953).
5. R. S. Mulliken, *J. Chim. Phys.*, **46**, 497, 675 (1949)
6. R. L. Hummel and K. Ruedenberg, *J. Phys. Chem.*, **66**, 2334 (1962).
7. M. D. Newton, F. P. Boer and W. N. Lipscomb, *J. Amer. Chem. Soc.*, **88**, 2353, 2361, 2367 (1966).
8. B. J. Nicholson, *Advan. Chem. Phys.*, **18**, 249 (1970).
9. L. Burnelle, *J. Chem. Phys.*, **35**, 311 (1961).
10. J. P. Dahl and C. J. Ballhausen, *Advan. Quantum Chem.*, **4**, 170 (1968).
11. L. Oleari, G. de Michelis and L. di Sipio, *Mol. Phys.*, **10**, 111 (1966).
11a. C. A. L. Becker and J. P. Dahl, *Theoret. Chim. Acta*, **14**, 26 (1969).
11b. J. P. Dahl and H. Johansen, *Theoret. Chim. Acta*, **11**, 8, 26 (1968).
11c. G. C. Allen and D. W. Clack, *J. Chem. Soc.*, A, 2668 (1970).
11d. D. W. Clack and M. S. Farrimond, *J. Chem. Soc.*, A, 299 (1971).
12. J. N. Murrell and J. E. Williams, unpublished results.
13. E. Switkes, R. M. Stevens and W. N. Lipscomb, *J. Chem. Phys.*, **51**, 5229 (1969).
14. W. E. Palke and W. N. Lipscomb, *J. Amer. Chem. Soc.*, **88**, 2384 (1966).
15. K. R. Roby and O. Sinanôglu, *Internat. J. Quantum Chem*, 3s, 223 (1969).

Appendix 1

The variation and perturbation methods

The variation theorem may be stated as follows: the energy calculated from an approximate wave function is always greater than the exact energy of the lowest state of the same symmetry. Let $\mathcal{H}$ be the Hamiltonian and Ψ the wave function, then from the theorem we can write†

$$E = \frac{\displaystyle\int \Psi^* \mathcal{H} \Psi \, d\tau}{\displaystyle\int \Psi^* \Psi \, d\tau} \equiv \frac{\langle \Psi | \mathcal{H} | \Psi \rangle}{\langle \Psi | \Psi \rangle} \geq E_0 \qquad \text{(A1.1)}$$

Ψ can always be expanded in the set of eigenfunctions of $\mathcal{H}$

$$\Psi = \sum_i c_i \Psi_i \qquad \text{(A1.2)}$$

whence

$$E = \frac{\displaystyle\sum_i \sum_j c_i c_j \langle \Psi_i | \mathcal{H} | \Psi_j \rangle}{\displaystyle\sum_i \sum_j c_i c_j \langle \Psi_i | \Psi_j \rangle} = \frac{\displaystyle\sum_i \sum_j c_i c_j \mathcal{H}_{ij}}{\displaystyle\sum_i \sum_j c_i c_j S_{ij}} . \qquad \text{(A1.3)}$$

However, if we use the fact that the eigenfunctions of $\mathcal{H}$ are an orthonormal set, $S_{ij} = \delta_{ij}$, and that their matrix elements of $\mathcal{H}$ are diagonal, $\mathcal{H}_{ij} = E_i \delta_{ij}$, (A1.3) takes the form

$$E = \frac{\displaystyle\sum_i c_i^2 \langle \Psi_i | \mathcal{H} | \Psi_i \rangle}{\displaystyle\sum_i c_i^2 \langle \Psi_i | \Psi_i \rangle} = \frac{\displaystyle\sum_i c_i^2 E_i}{\displaystyle\sum_i c_i^2} . \qquad \text{(A1.4)}$$

It follows that

$$\sum_i c_i^2 (E_i - E) = 0. \qquad \text{(A1.5)}$$

† Throughout the book we have assumed that wave functions have been chosen to be in real form. If complex functions are used integrals have the form $\int \Psi^* \mathcal{H} \Psi \, d\tau$, etc.

158

This identity can only hold if at least one factor $(E_i - E)$ is negative, and if Ψ_0 is the lowest energy state that enters into the expansion (A1.2), then $E_0 - E$ is the most negative factor. It follows that E must be greater than E_0.

The variation theorem is the basis of the most widely used mathematical technique in molecular quantum mechanics, which is the method of linear variation of constants. We write an approximate wave function Ψ as an expansion in a set of functions θ_i, which may or may not be eigenfunctions of some operator, and may or may not be orthonormal.

$$\Psi = \sum_i c_i \theta_i. \tag{A1.6}$$

The expectation value of the energy is given by an expression like (A1.3), which we can put in the form

$$\sum_i \sum_j c_i c_j (\mathscr{H}_{ij} - ES_{ij}) = 0. \tag{A1.7}$$

If the coefficient c_i is varied we have, after differentiating (A1.7),

$$\sum_j c_j (\mathscr{H}_{ij} - ES_{ij}) - \sum_i \sum_j c_i c_j S_{ij} (\partial E / \partial c_i) = 0. \tag{A1.8}$$

If we make use of the variation theorem and require that the coefficients be chosen in such a way that E is a minimum, then $(\partial E / \partial c_i) = 0$ for all c_i, hence

$$\sum_j c_j (\mathscr{H}_{ij} - ES_{ij}) = 0 \tag{A1.9}$$

for all i.

The set of simultaneous equations (A1.9) has a non trivial solution only if

$$|\mathscr{H}_{ij} - ES_{ij}| = 0 \tag{A1.10}$$

and on expanding this determinant we get a polynomial in E, whose order is the number of functions in the expansion (A1.6). The lowest solution of (A1.10) is the energy of the function Ψ which best approximates to the lowest eigenvalue of $\mathscr{H}$ for a state of the same symmetry. It can be shown that the other solutions of (A1.10) are lower bounds to successive higher energy states of $\mathscr{H}$ for that symmetry.

The equations (A1.9) are called the secular equations and (A1.10) is the secular determinant. These equations are met in the LCAO method where the functions θ_i are atomic orbitals used to expand a molecular orbital. They also occur in the method of configuration interaction, where the θ_i are many-electron wave functions (Slater determinants) which correspond to different ways of assigning electrons to a set of molecular orbitals. The resulting wave function Ψ will be a linear combination of wave functions associated with different electron configurations.

In perturbation theory the Hamiltonian is divided into two terms

$$\mathcal{H} = \mathcal{H}^0 + \lambda\mathcal{H}' \tag{A1.11}$$

where $\mathcal{H}^0$ is an operator whose eigenfunctions and eigenvalues are assumed to be known. The eigenfunctions and eigenvalues of $\mathcal{H}$ are then expanded as a power series in the parameter λ.

$$\Psi = \Psi_0 + \lambda\Psi^1 + \lambda^2\Psi^2 + \cdots$$
$$E = E_0 + \lambda E^1 + \lambda^2 E^2 + \cdots. \tag{A1.12}$$

If (A1.12) is substituted into the equation

$$(\mathcal{H} - E)\Psi = 0 \tag{A1.13}$$

then the collected terms for each order of λ must be individually zero. This leads to the following set of equations

$$\lambda^0: \qquad\qquad (\mathcal{H}^0 - E_0)\Psi_0 = 0$$
$$\lambda^1: \qquad\quad (\mathcal{H}^0 - E_0)\Psi^1 + (\mathcal{H}' - E^1)\Psi_0 = 0 \tag{A1.14}$$
$$\lambda^2: \quad (\mathcal{H}^0 - E_0)\Psi^2 + (\mathcal{H}' - E^1)\Psi^1 + (-E^2)\Psi_0 = 0 \text{ etc.}$$

If the function ψ is assumed to be normalized to all orders of λ, then we have

$$\lambda^1: \qquad\qquad 2\langle\Psi_0|\Psi^1\rangle = 0$$
$$\lambda^2: \quad \langle\Psi^1|\Psi^1\rangle + 2\langle\Psi_0|\Psi^2\rangle = 0 \text{ etc.} \tag{A1.15}$$

If we multiply the second equation in (A1.14) by Ψ_0 and integrate, and make use of the first equation in (A1.15), we have

$$\langle\Psi_0|\mathcal{H}' - E^1|\Psi_0\rangle = 0 \tag{A1.16}$$

which defines the first-order correction to the energy as the expectation value of Ψ'' for the operator $\mathcal{H}_0$

$$E^1 = \langle\Psi_0|\mathcal{H}'|\Psi_0\rangle. \tag{A1.17}$$

There are several different ways of getting precise expressions for the higher order energies and the expansion terms of the wave function. For a full account of these, together with other techniques of perturbation theory we refer the reader to an article by Hirschfelder, Byers Brown and Epstein.[1]

We give here the most common expressions, which are obtained by the Rayleigh–Schrödinger method in which the terms Ψ'^n are expanded in eigenfunctions of the operator $\mathscr{H}^0$. For Ψ'^1 we can write

$$\Psi'^1 = \sum_{i \neq 0} c_i \Psi'_i \qquad (A1.18)$$

and on substituting this into the second equation of (A1.14), we obtain

$$\sum_i' c_i(\mathscr{H}^0 - E_0)\Psi'_i + (\mathscr{H}' - E^1)\Psi'_0 = 0. \qquad (A1.19)$$

Multiplying (A1.19) by Ψ'_j and integrating gives

$$c_j = \frac{\langle \Psi'_j | \mathscr{H}' | \Psi'_0 \rangle}{E_0 - E_j} \qquad (A1.20)$$

where we have used the fact that the matrix of $\mathscr{H}^0$ is diagonal in its eigenfunctions.

On substituting (A1.18) into the third equation of (A1.14), multiplying by Ψ'_0 and integrating gives the second order energy as a sum-over-states expression

$$E^2 = \frac{\sum_i' \langle \Psi'_0 | \mathscr{H}' | \Psi'_i \rangle^2}{E_0 - E_i} \qquad (A1.21)$$

References

1. J. O. Hirschfelder, W. Byers Brown and S. T. Epstein, *Advan. Quantum Chem.*, **1**, 255 (1964).

The derivation of the Hartree–Fock equations

There are several ways of deriving the Hartree–Fock equations, but they are all based on the condition that a Hartree–Fock wave function is defined as the best fully antisymmetrized wave function which can be constructed from n spin orbitals (n being the number of electrons). The variation theorem provides the criterion of best, as being the wave function which has the minimum energy.

The concept of an orbital (atomic or molecular) is associated with the idea that an electron has a wave function ψ which defines, through ψ^2, its distribution in space. This wave function is determined by the attractive potential energy of the electron and the nuclei and the average repulsive potential energy with the other electrons in the molecule. However, ψ is not a complete description of the wave function for the electron because an electron has a spin. Consequently to describe the full wave function of the electron we must multiply its space function ψ by either of the possible spin functions (α or β). The result, $\psi\alpha$ or $\psi\beta$ is called a spinorbital. We will define an orthonormal set of these as $\chi_a, \chi_b, \ldots$. Any two may be orthogonal either because their space parts are orthogonal or because they have different spin functions.

If we have a system of two electrons then we can write a wave function of the pair as the product

$$\chi_a(1)\chi_b(2). \tag{A2.1}$$

However, this is not an acceptable form for an electron wave function, since we know that such functions must be antisymmetric to exchange of the coordinates of the two electrons. We can achieve this condition by taking the combination

$$\Psi = \sqrt{\tfrac{1}{2}}\{\chi_a(1)\chi_b(2) - \chi_a(2)\chi_b(1)\} \tag{A2.2}$$

in which the factor $\sqrt{\frac{1}{2}}$ is introduced as a normalization constant so that

$$\int \Psi'^2 \, d\tau = 1 \tag{A2.3}$$

or, in other words, there is a unit probability of finding both electrons somewhere in space. The volume element $d\tau$ is in the complete space of the position and spin coordinates of both electrons.

The Pauli exclusion principle can be formulated as 'only one electron can occupy a spin-orbital,' because it follows from (A2.2) that no anti-symmetric wave function like (A2.2) can be constructed if $\chi_a = \chi_b$.

The function (A2.2) is more conveniently written as a determinant

$$\Psi' = \sqrt{\tfrac{1}{2}} \begin{vmatrix} \chi_a(1) & \chi_b(1) \\ \chi_a(2) & \chi_b(2) \end{vmatrix} \tag{A2.4}$$

because this form is easily generalized to many electrons as follows

$$\Psi = (n!)^{-\frac{1}{2}} \begin{vmatrix} \chi_a(1) & \chi_b(1) & \cdots & \chi_r(1) \\ \chi_a(2) & \chi_b(2) & \cdots & \chi_r(2) \\ \hdotsfor{4} \\ \chi_a(n) & \chi_b(n) & \cdots & \chi_r(n) \end{vmatrix} \equiv |\chi_a \ \ \chi_b \ \ \cdots \ \ \chi_r|. \tag{A2.5}$$

$n!$ is the number of products obtained by expanding the determinant and since each product is orthogonal (because the χ's are orthogonal) then $(n!)^{-\frac{1}{2}}$ is the correct normalization constant. The antisymmetry condition is satisfied by such a function because determinants change sign if two rows are exchanged. The Pauli principle follows because a determinant is zero if two columns are identical.

The Hartree–Fock wave function is usually defined as that function (A2.5) which gives the lowest energy. For open-shell molecules a definition is occasionally used in which a limited number of such functions with different spin arrangements are chosen so as to make Ψ an eigenfunction of the total spin operator S^2. We shall not discuss such a definition.

In order to obtain the energies for many electron wave functions, we will first of all evaluate the energy of the two-electron wave function. We use the Hamiltonian

$$\mathcal{H} = \sum_i H^c(i) + \sum_{i<j} r_{ij}^{-1} \tag{A2.6}$$

where

$$H^c(i) = -\tfrac{1}{2}\nabla_i^2 - \sum_A Z_A r_{ai}^{-1}. \tag{A2.7}$$

We are here using atomic units (e, $\hbar$, $m = 1$), and have taken the system to contain nuclei of charge Z_A. The energy of the function (A2.4) is then

$$E = \tfrac{1}{2} \iint \{\chi_a(1)\chi_b(2) - \chi_a(2)\chi_b(1)\}\{\mathbf{H}^c(1) + \mathbf{H}^c(2) + r_{12}^{-1}\}$$
$$\times \{\chi_a(1)\chi_b(2) - \chi_a(2)\chi_b(1)\}\ d\tau_1\ d\tau_2. \quad (A2.8)$$

Because of the equivalence of the electrons we can collect the terms as follows

$$E = \int \chi_a(1)\mathbf{H}^c(1)\chi_a(1)\ d\tau_1 \int \chi_b(2)\chi_b(2)\ d\tau_2$$
$$+ \int \chi_b(1)\mathbf{H}^c(1)\chi_b(1)\ d\tau_1 \int \chi_a(2)\chi_a(2)\ d\tau_2$$
$$- 2\int \chi_a(1)\mathbf{H}^c(1)\chi_b(1)\ d\tau_1 \int \chi_a(2)\chi_b(2)\ d\tau_2$$
$$+ \iint \chi_a(1)\chi_b(2)r_{12}^{-1}\chi_a(1)\chi_b(2)\ d\tau_1\ d\tau_2$$
$$- \iint \chi_a(1)\chi_b(2)r_{12}^{-1}\chi_b(1)\chi_a(2)\ d\tau_1\ d\tau_2. \quad (A2.9)$$

This expression can be simplified further, because the conditions of orthogonality of two spin-orbitals will make the third integral zero. Also we use normalized functions so that

$$\int \chi_a\chi_a\ d\tau = 1 \quad (A2.10)$$

The resulting expression we write

$$E = H_{aa}{}^c + H_{bb}{}^c + (\chi_a\chi_a \mid \chi_b\chi_b) - (\chi_a\chi_b \mid \chi_a\chi_b) \quad (A2.11)$$

where

$$H_{aa}{}^c \equiv \int \chi_a(1)\mathbf{H}^c(1)\chi_a(1)\ d\tau_1 \quad (A2.12)$$

and

$$(\chi_a\chi_b \mid \chi_a\chi_b) \equiv \iint \chi_a(1)\chi_b(2)r_{12}^{-1}\chi_b(1)\chi_a(2)\ d\tau_1\ d\tau_2. \quad (A2.13)$$

The integral $(\chi_a\chi_a \mid \chi_b\chi_b)$ is called a coulomb integral, and $(\chi_a\chi_b \mid \chi_a\chi_b)$ an exchange integral. For the many electron wave functions (A2.5), expression (A2.11) may be extended in an obvious way to give

$$E = \sum_r H_{rr}{}^c + \sum_{r>s} \sum [(\chi_r\chi_r \mid \chi_s\chi_s) - (\chi_r\chi_s \mid \chi_r\chi_s)]. \quad (A2.14)$$

One way to derive the Hartree–Fock equations is to consider the variation of E, as given by (A2.14), with respect to changes in the functions χ_r and χ_s subject to the restriction that these functions remain orthonormal. In this way a differential equation can be derived of the form

$$\mathbf{F}\chi = \epsilon\chi \qquad (A2.15)$$

where $\mathbf{F}$ is the Hartree–Fock operator. Unfortunately operator equations like (A2.15) are too difficult to solve directly except for atoms which have a high symmetry. For molecules the equations are always solved by matrix methods. It is a fundamental theorem of quantum mechanics that the solutions of an equation like (A2.15) are orthogonal to one another. We therefore have

$$F_{at} \equiv \int \chi_a \mathbf{F}\chi_t \, d\tau = \int \chi_a \epsilon_t \chi_t \, d\tau = \epsilon_t \int \chi_a \chi_t \, d\tau = \epsilon_t \delta_{at} \qquad (A2.16)$$

where δ is the Kronecker delta which is zero for $a \neq t$ and unity for $a = t$.

We say that the matrix F_{at} is diagonal in the basis of eigenfunctions of the operator $\mathbf{F}$. The method we shall follow to derive the Hartree–Fock equation is to obtain directly a general expression for F_{at}, rather than for the operator $\mathbf{F}$.

A determinant can be expanded in the following manner

$$|(\chi_a + \lambda\chi_t)\chi_b \cdots \chi_r| = |\chi_a\chi_b \cdots \chi_r| + \lambda |\chi_t\chi_b \cdots \chi_r|. \qquad (A2.17)$$

We shall write these functions as

$$\Psi'' = \Psi + \lambda\Psi_{at} \qquad (A2.18)$$

where Ψ_{at} indicates the determinant obtained from Ψ by replacing χ_a by χ_t. We note that Ψ and Ψ_{at} are orthogonal because χ_a and χ_t are orthogonal. The expression for the energy of Ψ'' can be written down quite readily as

$$E' = \frac{\int \Psi'' \mathscr{H} \Psi'' \, d\tau}{\int \Psi'' \Psi'' \, d\tau} = \frac{\int \Psi \mathscr{H} \Psi \, d\tau + 2\lambda \int \Psi \mathscr{H} \Psi_{at} \, d\tau + \lambda^2 \int \Psi_{at} \mathscr{H} \Psi_{at} \, d\tau}{1 + \lambda^2}$$

$$(A2.19)$$

the normalization factor is introduced since Ψ'', unlike Ψ and Ψ_{at}, is not normalized.

For small values of λ expression (A2.19) reduces to

$$E' = E + 2\lambda \int \Psi \mathscr{H} \Psi_{at} \, d\tau \qquad (A2.20)$$

where E is the energy of Ψ. It follows from (A2.20) that it is always possible to find a value of λ such that $E' < E$ unless

$$\int \Psi \mathscr{H} \Psi_{at} \, d\tau = 0. \qquad (A2.21)$$

If Ψ is the Hartree–Fock wave function then it must have a lower energy than any function Ψ' which differs from it. This can only be possible if all integrals like (A2.21) are zero, we shall denote this integral by F_{at}, in recognition of the condition noted earlier that for Hartree–Fock orbitals F_{at} is a diagonal matrix.

The integral (A2.21) is evaluated rather easily if the simplifying rules for handling integrals between many-electron wave functions are known.[1,2] Expanding this integral we obtain

$$
\begin{aligned}
F_{at} &= \int |\chi_a \chi_b \cdots \chi_r| \, \mathscr{H} \, |\chi_t \chi_b \cdots \chi_r| \, d\tau \\
&= H_{at}{}^c + \sum_{k=a}^{r} [(\chi_a \chi_t | \chi_k \chi_k) - (\chi_a \chi_k | \chi_t \chi_k)]. \qquad (A2.22)
\end{aligned}
$$

This expression contains integrals similar to those occurring in (A2.14).

The integrals in (A2.22) are over both the position and spin coordinates of the electrons. However, since the operators H^c and $r_{12}{}^{-1}$ do not involve spin we can readily integrate over the spin coordinates. A difficulty here, however, is that the form adopted by F_{at} then depends upon the spin function actually used. We first consider the most important case which applies to the ground states of most molecules, namely when we have a closed shell of two electrons in each occupied molecular orbital. The resulting wave function is usually called a restricted Hartree–Fock function. It can be seen from (A2.22) that each term in F_{at} will be zero unless χ_a and χ_t have the same spin function. Let us call this spin α; then we have

$$\chi_a = \psi_a \alpha, \quad \chi_t = \psi_t \alpha. \qquad (A2.23)$$

If χ_k is also an α spin-orbital then the two-electron terms in (A2.22) give

$$(\psi_a \psi_t | \psi_k \psi_k) - (\psi_a \psi_k | \psi_t \psi_k). \qquad (A2.24)$$

If, however, χ_k has β spin then only the first term would remain. For a closed-shell molecule both $\psi_k \alpha$ and $\psi_k \beta$ are occupied, and we can therefore obtain an expression in which we sum over the occupied *orbitals* k rather than the spin orbitals

$$F_{at} = H_{at}{}^c + \sum_k [2(\psi_a \psi_t | \psi_k \psi_k) - (\psi_a \psi_k | \psi_t \psi_k)]. \qquad (A2.25)$$

This is the basic definition of the matrix elements of the Hartree–Fock operator for a closed-shell wave function.

When a basis ψ is obtained in which F_{at} is diagonal the diagonal elements may be defined as the orbital energies, thus

$$\epsilon_a = F_{aa} = H_{aa}^{\ c} + \sum_k [2(\psi_a\psi_a \mid \psi_k\psi_k) - (\psi_a\psi_k \mid \psi_a\psi_k)]$$

$$\equiv H_{aa}^{\ c} + \sum_k (2J_{ak} - K_{ak}), \tag{A2.26}$$

where J_{ak} and K_{ak} are called the coulomb and exchange integrals respectively. It is important to note that the total energy is not simply a sum of orbital energies. If we start with (A2.14) and integrate over the spin, then for the special case of a closed-shell wave function we have

$$E = \sum_r H_{rr}^{\ c} + \sum_r \sum_{s \neq r} (2J_{rs} - K_{rs}), \tag{A2.27}$$

where the summations are over occupied orbitals. However as $J_{rr} = K_{rr}$ we can include the terms $s = r$ in this second summation and obtain

$$E = \sum_r \left[2H_{rr}^{\ c} + \sum_s (2J_{rs} - K_{rs}) \right]. \tag{A2.28}$$

If the orbital energies are introduced from (A2.26), this becomes

$$E = \sum_r \left[2F_{rr} - \sum_s (2J_{rs} - K_{rs}) \right], \tag{A2.29}$$

and, finally, the repulsion energy of the nuclei has to be added.

If we use an LCAO expansion for the molecular orbitals

$$\psi_k = \sum_\mu c_{k\mu}\varphi_\mu, \tag{A2.30}$$

then the coefficients $c_{k\mu}$ which make $\mathbf{F}$ diagonal are obtained by solving

$$\sum_\mu c_{k\mu}(F_{\mu\nu} - ES_{\mu\nu}) = 0 \tag{A2.31}$$

for which a necessary condition is

$$|F_{\mu\nu} - ES_{\mu\nu}| = 0. \tag{A2.32}$$

The situation here is completely analogous to the method of finding the coefficients that make the Hückel matrix diagonal.

If we substitute (A2.30) into both sides of (A2.25) we obtain:

$$\sum_\mu \sum_\nu c_{a\mu}c_{t\nu}F_{\mu\nu} = \sum_\mu \sum_\nu c_{a\mu}c_{t\nu}H_{\mu\nu}^{\ c}$$
$$+ \sum_k \sum_\mu \sum_\nu \sum_\rho \sum_\sigma c_{a\mu}c_{t\nu}c_{k\rho}c_{k\sigma}$$
$$\times [2(\mu\nu \mid \rho\sigma) - (\mu\rho \mid \nu\sigma)]. \tag{A2.33}$$

By equating the terms in $c_{a\mu}c_{tv}$ on both sides of this expression we get

$$F_{\mu v} = H_{\mu v}^{\ c} + \sum_k \sum_\rho \sum_\sigma c_{k\rho}c_{k\sigma}[2(\mu v \mid \rho\sigma) - (\mu\rho \mid v\sigma)] \quad \text{(A2.34)}$$

and using the definition of the bond order

$$P_{\rho\sigma} = 2\sum_k c_{k\rho}c_{k\sigma} \quad \text{(A2.35)}$$

we obtain the required definition of the F-matrix elements over atomic orbitals

$$F_{\mu v} = H_{\mu v}^{\ c} + \sum_\rho \sum_\sigma P_{\rho\sigma}[(\mu v \mid \rho\sigma) - \tfrac{1}{2}(\mu\rho \mid v\sigma)]. \quad \text{(A2.36)}$$

This expression was our starting point for the equations governing the SCF orbitals in the semi-empirical theories we have described.

The SCF equations (A2.31) are cubic in the coefficients and can only be solved by an iterative method. The usual procedure is to assign an initial set of coefficients (normally these are the Hückel coefficients) and to calculate the elements of the bond-order matrix by (A2.35). The secular determinent (A2.32) is then solved and a new set of coefficients obtained from (A2.31). These are then used to repeat the cycle, and the process continued until the bond orders and charge densities obtained from two consecutive iterations agree to within the desired accuracy; the calculated wave functions are then said to be *self-consistent*.

A wave function like (A2.5) in which each spin-orbital has a different orbital function is called an unrestricted Hartree–Fock function. The matrix elements (A2.22) are again zero unless χ_a and χ_t have the same spin functions, so the F matrix may be decomposed into an α and a β block whose elements are

$$F_{at}^{\ \alpha} = H_{at}^{\ c} + \sum_{k(\alpha)} [(\chi_a\chi_t \mid \chi_k\chi_k) - (\chi_a\chi_k \mid \chi_t\chi_k)] + \sum_{k(\beta)} (\chi_a\chi_t \mid \chi_k\chi_k) \quad \text{(A2.37)}$$

$$F_{at}^{\ \beta} = H_{at}^{\ c} + \sum_{k(\beta)} [(\chi_a\chi_t \mid \chi_k\chi_k) - (\chi_a\chi_k \mid \chi_t\chi_k)] + \sum_{k(\alpha)} (\chi_a\chi_t \mid \chi_k\chi_k) \quad \text{(A2.38)}$$

where the summations over $k(\alpha)$ and $k(\beta)$ spin-orbitals have been separated.

When the spin coordinates are integrated out of these expressions and the LCAO expansion (A2.30) introduced, one obtains the secular equations analogous to (A2.31), but with separate equations for the α and β orbitals. Thus the coefficients of the α set are given by

$$\sum c_{k\mu}^{\ \alpha}(F_{\mu v}^{\ \alpha} - ES_{\mu v}) = 0 \quad \text{(A2.39)}$$

and

$$|F_{\mu v}^{\ \alpha} - ES_{\mu v}| = 0 \quad \text{(A2.40)}$$

where, by analogy with (A2.32) we have

$$F_{\mu\nu}{}^{\alpha} = H_{\mu\nu}{}^{c} + \sum_{\rho} \sum_{\sigma} \{P_{\rho\sigma}{}^{\alpha}[(\mu\nu \mid \rho\sigma) - (\mu\rho \mid \nu\sigma)] + P_{\rho\sigma}{}^{\beta}(\mu\nu \mid \rho\sigma)\}. \quad \text{(A2.41)}$$

We are here introducing separate bond orders for the α and β electrons:

$$P_{\rho\sigma}{}^{\alpha} = \sum_{k(\alpha)} c_{k\rho}c_{k\sigma}, \quad \text{(A2.42)}$$

$$P_{\rho\sigma} = P_{\rho\sigma}{}^{\alpha} + P_{\rho\sigma}{}^{\beta}. \quad \text{(A2.43)}$$

There will be a corresponding set of equations to determine the β orbitals. For full self-consistency both the α and β matrices must be diagonalized simultaneously because the two matrices are linked by having common bond orders. The iterative cycle therefore consists of an assumed set of coefficients which are used to construct and solve one set of secular equations (α say). The resulting improved α coefficients together with the assumed β set are then used to construct the secular equations for the β set, and these are solved. Thus the α and β sets are treated alternately until self-consistency is reached.

Expression (A2.36) for the restricted Hartree–Fock matrix elements were derived by Lennard-Jones,[3] Hall[4] and Roothaan.[5] The unrestricted Hartree–Fock method described here was suggested by Pople and Nesbet[6] and the LCAO expressions from the F-matrix elements were given by Brickstock and Pople.[7] Roothaan has described a more general treatment of open shells.[8]

References

1. J. N. Murrell, S. F. A. Kettle and J. M. Tedder, *Valence Theory*, Wiley, London, 1965.
2. H. Eyring, J. Walter and G. E. Kimball, *Quantum Chemistry*, Wiley, New York, 1944.
3. J. E. Lennard-Jones, *Proc. Roy. Soc. (London)*, **A198**, 1, 14 (1949).
4. G. G. Hall, *Proc. Roy. Soc. (London)*, **A205**, 541 (1951).
5. C. C. J. Roothaan, *Rev. Mod. Phys.* **23**, 69 (1951).
6. J. A. Pople and R. K. Nesbet, *J. Chem. Phys.*, **22**, 571 (1954).
7. A. Brickstock and J. A. Pople, *Trans. Faraday Soc.*, **50**, 901 (1954).
8. C. C. J. Roothaan, *Rev. Mod. Phys.*, **32**, 179 (1960).

Author index

171

Subject index